T0324887

# METHODS IN MOLECULAR BIOLOGY

*Series Editor*
**John M. Walker**
**School of Life and Medical Sciences**
**University of Hertfordshire**
**Hatfield, Hertfordshire, AL10 9AB, UK**

For further volumes:
http://www.springer.com/series/7651

# Plant Cytogenetics

## Methods and Protocols

Edited by

## Shahryar F. Kianian

*University of Minnesota, St. Paul, MN, USA*

## Penny M.A. Kianian

*University of Minnesota, St. Paul, MN, USA*

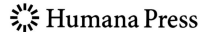

*Editors*
Shahryar F. Kianian
University of Minnesota
St. Paul, MN, USA

Penny M.A. Kianian
University of Minnesota
St. Paul, MN, USA

ISSN 1064-3745            ISSN 1940-6029   (electronic)
Methods in Molecular Biology
ISBN 978-1-4939-3620-5    ISBN 978-1-4939-3622-9   (eBook)
DOI 10.1007/978-1-4939-3622-9

Library of Congress Control Number: 2016941227

Photo Credit: Dr. Adam J. Lukaszewski, Dr. D. Kopecky, and Dr. G. Linc.

Printed on acid-free paper

This Humana Press imprint is published by Springer Nature
The registered company is Springer Science+Business Media LLC New York

# Foreword: The Modern Cytogenetics Tool Box—A Picture Is Still Worth a Thousand Words

Cytogenetics is a stepping stone toward a full understanding of genetics. The report of in situ hybridization by Pardue and Gall in 1969 introduced a new era combining cytogenetics and molecular biology. An important feature was that allelic variation was not required for placing important genes to chromosome—and their place could be directly visualized with the light microscope. The narrative of this book takes the reader from the neoclassical to modern technologies for visualizing chromosomes, chromosome segments, and DNA strands. Many of the techniques require modest equipment and other resources. The book addresses complex situations including polyploidy in species such as oat and wheat. The chapters in this book are clearly written and provide useful protocols and the appropriate references. The description of C banding shows how such a slight modification of standard cytogenetic techniques can provide previously inaccessible information relative to deletions, translocations, and other chromosome structural changes and can be used in place of more extensive—and expensive—molecular technologies. It is an easy jump from C banding to the use of genomic in situ hybridization (GISH) and fluorescence in situ hybridization (FISH). Certainly, these techniques show that "a picture is worth a thousand words". From distinguishing ancestral genomes to placing single-copy sequences relative to each other has yielded important insights in many cases, such as detecting gaps between BACs. Fiber FISH further advanced resolution by identifying each individual chromosome and localizing sequences on individual chromatin fibers. This technique is now being enhanced through tyramide signal amplification (TSA-FISH) of low copy sequences which increases the sensitivity of FISH perhaps by 1000-fold.

Recombination is the basis of genetic compositions selected by the breeder. Further understanding of the frequency and distribution of recombination points could become the basis of major advances in breeding. Genes exist that control homologous and homoeologous recombination. The use of such genes affecting recombination can lead to unique combinations of chromosome segments: ideas for achieving such goals are described in this book.

Analysis of genomes is sometimes made simpler by using radiation to cause chromosome breakages. The creation of Radiation Hybrids has provided not only information on physical linkages via high-resolution physical maps but also many new cytogenetic stocks for use in genetic experimentation. Further resolution can be achieved through cutting chromosome fibers by restriction enzymes and the placement of sequences by optical mapping (also described in this book).

The flow sorting of chromosomes simplifies genomic analysis by separating out specific chromosomes or chromosome segments. This approach was important for the sequencing of the wheat genome in that BAC libraries were made from individual flow-sorted chromosomes or segments almost free of organellar DNA. A complete set of chromosome arm-specific BAC libraries were constructed for wheat. Chromosome microdissection is another method described in this book for reducing the complexity of subsequent genome analysis. Isolated interphase nuclei also avoid some of the complications of tissue isolation that may provide an unwanted mixture of cell types. This approach can be followed by subsequent immunolocalization of proteins to show spatial and temporal relationships, such as the described RAD51 involved in the repair of double-stranded breaks. Epigenomic modifications of the genome are now well-established events and also can be increasingly understood by chromosome visualization techniques involving chromatin immunoprecipitation techniques.

The realm of cytogenetics—as described in this book—continues to expand and provides clear insights for genomic analyses.

*St. Paul, MN, USA*                                                        *Ronald L. Phillips*

# Preface

Cytogenetic studies have contributed greatly to our understanding of genetics, biology, reproduction, and evolution. From early studies in basic chromosome behavior, the field has expanded enabling whole genome analysis to the manipulation of chromosomes and their organization. This book covers a range of methods used in cytogenetics, beginning with basic analysis of chromosomes and visualizing gene locations (Chapters 1–6), to manipulating and dissecting chromosomes (Chapters 7–12), and then focusing on less-understood features of chromosomes such as recombination initiation sites and epigenomic marks (Chapters 13–15). The methods described are detailed and built on each other, assisting those new to the field a comprehensive platform to support their research endeavor, while introducing advanced techniques to experienced researchers. We hope this book starts you on an adventure into the field of cytogenetics, while you discover the wonderment of the complexity of nature and beauty of the biologically important chromosomes through your microscope eyepiece.

*St. Paul, MN, USA*

*Shahryar F. Kianian*
*Penny M.A. Kianian*

# Contents

# Contributors

JAN BARTOŠ • *Centre of the Region Haná for Biotechnological and Agricultural Research, Institute of Experimental Botany, Olomouc, Czech Republic*

XIWEN CAI • *Department of Plant Sciences, North Dakota State University, Fargo, ND, USA*

PETR CÁPAL • *Centre of the Region Haná for Biotechnological and Agricultural Research, Institute of Experimental Botany, Olomouc, Czech Republic*

CHANGBIN CHEN • *Department of Horticultural Science, University of Minnesota, St. Paul, MN, USA*

JARMILA ČÍHALÍKOVÁ • *Centre of the Region Haná for Biotechnological and Agricultural Research, Institute of Experimental Botany, Olomouc, Czech Republic*

KARIN R. DEAL • *Department of Plant Sciences, University of California, Davis, CA, USA*

DARYNA DECHYEVA • *Institute of Botany, Technische Universität Dresden, Dresden, Germany*

CHUAN-LIANG DENG • *Henan Normal University, Xinxiang, China*

JAROSLAV DOLEŽEL • *Centre of the Region Haná for Biotechnological and Agricultural Research, Institute of Experimental Botany, Olomouc, Czech Republic*

STEFANIE DUKOWIC-SCHULZE • *Department of Horticultural Science, University of Minnesota, St. Paul, MN, USA*

ESTHER FERRER • *Department of Biomedicine and Biotechnology, University of Alcalá, Alcalá de Henares, Madrid, Spain*

ARACELI FOMINAYA • *Department of Biomedicine and Biotechnology, University of Alcalá, Alcalá de Henares, Madrid, Spain*

JUAN M. GONZÁLEZ • *Department of Biomedicine and Biotechnology, University of Alcalá, Alcalá de Henares, Madrid, Spain*

ANTHONY HARRIS • *Department of Horticultural Science, University of Minnesota, St. Paul, MN, USA*

ALEX R. HASTIE • *BioNano Genomics, San Diego, CA, USA*

YAN HE • *National Maize Improvement Center of China, Beijing Key Laboratory of Crop Genetic Improvement, China Agricultural University, Beijing, China*

JUSTIN B. HEGSTAD • *Department of Plant Sciences, North Dakota State University, Fargo, ND, USA*

ZAN-MIN HU • *Institute of Genetics and Developmental Biology, Chinese Academy of Sciences, Beijing, China*

CHAO-CHIEN JAN • *USDA-ARS, Sunflower and Plant Biology Research Unit, Northern Crop Science Laboratory, Fargo, ND, USA*

ERIC N. JELLEN • *Department of Plant and Wildlife Sciences, Brigham Young University, Provo, UT, USA*

MIROSLAVA KARAFIÁTOVÁ • *Centre of the Region Haná for Biotechnological and Agricultural Research, Institute of Experimental Botany, Olomouc, Czech Republic*

PENNY M.A. KIANIAN • *Department of Horticultural Science, University of Minnesota, St. Paul, MN, USA*

SHAHRYAR F. KIANIAN • *USDA-ARS, Cereal Disease Laboratory, Department of Plant Pathology, University of Minnesota, St. Paul, MN, USA*

MARIE KUBALÁKOVÁ • *Centre of the Region Haná for Biotechnological and Agricultural Research, Institute of Experimental Botany, Olomouc, Czech Republic*

KATIE L. LIBERATORE • *USDA-ARS, Cereal Disease Laboratory, Department of Plant Pathology, University of Minnesota, St. Paul, MN, USA*

ZHAO LIU • *Department of Plant Sciences, North Dakota State University, Fargo, ND, USA*

YOLANDA LOARCE • *Department of Biomedicine and Biotechnology, University of Alcalá, Alcalá de Henares, Madrid, Spain*

ADAM J. LUKASZEWSKI • *Department of Botany and Plant Sciences, University of California, Riverside, CA, USA*

MING-CHENG LUO • *Department of Plant Sciences, University of California, Davis, CA, USA*

MARISA E. MILLER • *Department of Horticultural Science, University of Minnesota, St. Paul, MN, USA*

ARMOND MURRAY • *Department of Plant Sciences, University of California, Davis, CA, USA*

ZHIXIA NIU • *USDA-ARS, Cereal Crops Research Unit, Northern Crop Science Laboratory, Fargo, ND, USA*

WOJCIECH P. PAWLOWSKI • *Department of Plant Breeding and Genetics, School of Integrative Plant Science, Cornell University, Ithaca, NY, USA*

ALI PENDLE • *Department of Cell and Developmental Biology, John Innes Centre, Norwich, UK*

GERNOT PRESTING • *Department of Molecular Biosciences and Bioengineering, University of Hawaii at Manoa, Honolulu, HI, USA*

HENRY SADOWSKI • *BioNano Genomics, San Diego, CA, USA*

JAN ŠAFÁŘ • *Centre of the Region Haná for Biotechnological and Agricultural Research, Institute of Experimental Botany, Olomouc, Czech Republic*

MICHAEL SAGHBINI • *BioNano Genomics, San Diego, CA, USA*

THOMAS SCHMIDT • *Institute of Botany, Technische Universität Dresden, Dresden, Germany*

PETER SHAW • *Department of Cell and Developmental Biology, John Innes Centre, Norwich, UK*

HANA ŠIMKOVÁ • *Centre of the Region Haná for Biotechnological and Agricultural Research, Institute of Experimental Botany, Olomouc, Czech Republic*

WILL STEDMAN • *BioNano Genomics, San Diego, CA, USA*

QI SUN • *Institute of Biotechnology, Biotechnology Resource Center, Cornell University, Ithaca, NY, USA*

JAN VRÁNA • *Centre of the Region Haná for Biotechnological and Agricultural Research, Institute of Experimental Botany, Olomouc, Czech Republic*

MINGHUI WANG • *Institute of Biotechnology, Biotechnology Resource Center and Section of Plant Biology in School of Integrative Plant Science, Cornell University, Ithaca, NY, USA*

ZIDIAN XIE • *Department of Molecular Biosciences and Bioengineering, University of Hawaii at Manoa, Honolulu, HI, USA*

STEVEN S. XU • *USDA-ARS, Cereal Crops Research Unit, Northern Crop Science Laboratory, Fargo, ND, USA*

YING-XIN ZHANG • *Institute of Genetics and Developmental Biology, Chinese Academy of Sciences, Beijing, China; Graduate University of Chinese Academy of Sciences, Beijing, P. R. China*

QIJUN ZHANG • *Department of Plant Sciences, North Dakota State University, Fargo, ND, USA*

TINGTING ZHU • *Department of Plant Sciences, University of California, Davis, CA, USA*

# Chapter 1

# C-Banding of Plant Chromosomes

## Eric N. Jellen

## Abstract

C-banding is used to differentially stain metaphase chromosomes in organisms having appreciable amounts of constitutive heterochromatin. Its primary benefits are that it is an inexpensive and a relatively fast method of identifying individual chromosomes and morphological or karyotypic variation, including large chromosomal rearrangements and aneuploidies. We currently employ this technique with considerable effect in genome analysis of oat (*Avena sativa*) and related grass species, though it has been most extensively used for chromosome analysis of wheat (*Triticum aestivum*) and its relatives of the Triticeae.

**Key words** C-banding, Oats, *Avena*, Triticeae pooideae, Karyotyping

## 1 Introduction

The C-banding technique has been used extensively since the 1970s for cytogenetic studies of plant morphology, most notably in the Triticeae (Fig. 1), and its use has persisted through the era of molecular cytogenetics due to its relatively low cost and rapidity (for comprehensive review, *see* ref. 1). The first reports were published in 1973 in *Allium* [2–4], *Trillium, Vicia, Fritillaria*, and *Scilla* [5], and in the Triticeae for rye [6]. For example, C-banding is a more practical method than in situ hybridization for identifying gross morphological or karyotypic variation (Fig. 2) in large numbers of samples. It is also a much more accessible technique for scientists in the developing world, since C-banding includes no requirement for fluorescent microscopy, DNA labeling and hybridization reagents, or even computer-based imaging systems.

The basic C-banding method involves a series of chemical treatment steps of metaphase chromosome preparations affixed to standard microscope slides. Typically, metaphase chromosomes are accumulated through pretreatment with mitotic spindle inhibitors such as colchicine, colcemid (demecolcine), amiprophos-methyl (APM), 8-hydroxyquinoline, monobromonaphthalene, trifluralin

Shahryar F. Kianian and Penny M.A. Kianian (eds.), *Plant Cytogenetics: Methods and Protocols*,
Methods in Molecular Biology, vol. 1429, DOI 10.1007/978-1-4939-3622-9_1,

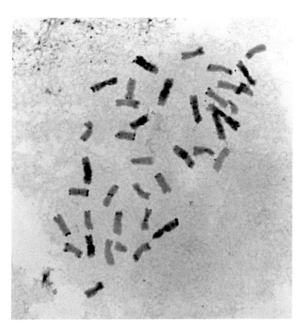

**Fig. 1** C-banded somatic metaphase chromosomes of *Avena sativa*

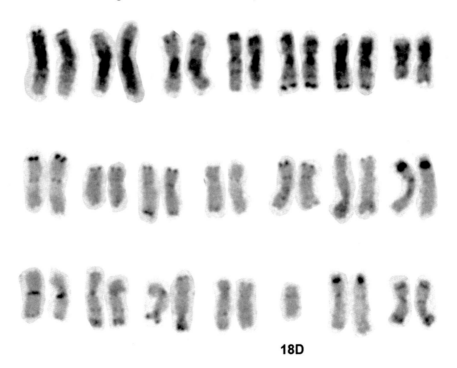

**18D**

**Fig. 2** C-banding karyotype of an *Avena sativa* plant monosomic for chromosome 18D (labeled)

(Treflan), ice water, or nitrous oxide gas. The chemical treatment steps of the prepared slides consist of an initial depurination wash in hot, dilute HCl, a wash in a concentrated alkaline solution at room temperature to denature the chromatin, a controlled renaturation wash in saline sodium citrate (SSC), and then staining in a phosphate-buffered eosin methylene blue-based stain.

The basic staining procedure was a modification from Giraldez et al. [7] and was first applied to *Avena* chromosomes in 1993 [8]. The inclusion of nitrous oxide gas treatment to arrest metaphase root-tip chromosomes follows the procedure of Kato [9].

## 2 Materials

Solutions do not need to be prepared with ultrapure water, as long as they are cycled through within a few months and are stored in the refrigerator.

### 2.1 Solutions

1. Farmer's Solution: 3:1 95 % ethanol: glacial acetic acid.
2. 45 % (v/v) acetic acid.
3. 100 % ethanol.
4. 20× SSC stock: 3 M NaCl, 0.3 M Na-citrate pH 7.0, adjusting pH with HCl and 5 M EDTA, if necessary.
5. 2× SSC is freshly prepared from a 20× SSC.
6. Giemsa stain solution: 1/15 M phosphate buffer [1:3 ($KH_2PO_4$:$Na_2HPO_4$)]. Geisma can be stored in tightly sealed bottles at room temperature (*see* **Note 1**).
7. Barium hydroxide: barium hydroxide octahydrate can be stored in tightly sealed bottles at room temperature. Solution is made by dissolving 300 g of $Ba(OH)_2 \cdot 8H_2O$ in 500 mL of distilled water. As long as crystals remain at the bottom of the flask, distilled water can be added to increase the volume of the saturated solution.
8. 0.2 M HCl: made freshly each time from a 1 M HCl stock.
9. Xylene.

### 2.2 Materials

1. Nitric oxide gas ($N_2O$).
2. Toluene-based mounting medium (e.g., Cytoseal).
3. Cover slip (#1 thickness).
4. Slide jars or trays for slide treatments.
5. Light microscope with phase contrast capability.

## 3 Methods

### 3.1 Root-Tip Metaphase Chromosome Pretreatment

1. Germinate seeds, embryo side up, on filter paper in petri plates at 19–21 °C in distilled water.
2. Excise root tips when 1–2 cm long, and treat with 130–150 psi $N_2O$ gas for 1.5–3 h.
3. Fix in Farmer's Solution overnight at room temperature.

### 3.2 Chromosome Preparations

1. Make a preparation of mitotic chromosomes in 45 % acetic acid. You can tap the cover slip over the root tip with a pencil tip while holding one edge of the cover slip stable using two fingers to break up the meristem. Another way to do it is to microdissect the root tip and squeeze or scrape out the meristematic cells and then discard the remainder of the root tip. If the mitotic index is high, either method will work fine.

   IMPORTANT: Do not use acetocarmine! Chromosome preparations can only be inspected using phase contrast microscopy at this step.

2. Heat the bottom of the slide gently using the open flame of an alcohol burner until it is hot to the touch. If it boils, remake the preparation.

3. Freeze slide on dry ice 5 min, pry off the cover slip using the edge of a razor blade, dehydrate in 100 % ethanol for at least 2 h, and then air-dry.

### 3.3 C-Banding Protocol

1. Treat slides with 0.2 M HCl, 60 °C for 2.5 min *precisely* (*see* **Note 2**), rinse, let slides drain *but not* totally dry.

2. Soak slides in saturated $BaOH_2$ at room temperature for approximately 7 min, and replace barium hydroxide completely with gently running warm (not hot) tap water (*see* **Note 3**). Then remove the slides and let them drain *but not* totally dry.

3. Soak slides in 2× SSC at 60 °C, for approx. 40 min, then shake off solution gently.

4. Stain slides in Giemsa stain solution (*see* **Note 1**). To check staining progress after 10 min, carefully remove slides to avoid iridescent filmy precipitate on the stain surface and rinse in two changes of tap water and then drain excess water. Slides can be inspected under a cover slip while wet and then float cover slip off in water (*see* **Note 4**).

5. Soak slides for a few minutes in xylene; mount using a small drop of mounting medium under a cover slip (#1 thickness) (*see* **Note 5**).

6. Observed slides using light microscope. Dark C-bands correspond to heterochromatic regions of the chromosomes. In plants, these bands are typically found at the centromeric and telomeric regions of the chromosome.

## 4  Notes

1. The Giemsa stain source is critical; I have used prepared liquid Giemsa from Sigma-Aldrich (#GS500) for many years, and the quality has been consistently excellent up to approximately 6 months past the expiration date on the bottle. In the past, I

have also used Leishman and Wright stains, but would not recommend them now.

2. IMPORTANT: Timing is critical! If not treated for a sufficient length of time, the staining will be overly blue; too long of a treatment results in pale pink staining.

3. For $BaOH_2$ treatment, break surface film on the $BaOH_2$ solution with the edge of the slide to avoid depositing a layer of the precipitate film over the area containing the chromosome preparation! Slides should not be lifted directly out of the solution; instead, displace the $BaOH_2$ solution with a slow stream of warm tap water for 1–2 min to avoid deposition of barium carbonate precipitate onto the slide surface. Timing in $BaOH_2$ solution is not critical, $\pm 1$ min.

4. If chromosomes are too purple, increase amount of time in the HCl step. If chromosomes are too pink, decrease time in HCl.

5. Mounting medium should be clear and colorless so that it will not crack or discolor with age (e.g., Cytoseal). Do not use Euparal media.

## References

1. Zoshchuk NV, Badaeva ED, Zelenin AV (2003) History of modern chromosomal analysis. Differential staining of plant chromosomes. Russ J Dev Biol 34:1–13

2. Greilhuber J (1972) Differentielle Heterochromatin farbung und Darstellung von Schraubenbau sowie Subchromatiden an pflanzlichen somatischen Chromosomen in der Meta- und Anaphase. Osterr Bot Z 121:1–11

3. Stack SM, Clarke CR (1973) Pericentromeric chromosome banding in higher plants. Can J Genet Cytol 15:367–369

4. Stack SM, Clarke CR (1973) Differential Giemsa staining of the telomeres of *Allium cepa* chromosomes: observations related to chromosome pairing. Can J Genet Cytol 15: 619–624

5. Schweizer D (1973) Differential staining of plant chromosomes with Giemsa. Chromosoma 40:307–320

6. Sharma NP, Natarajan AT (1973) Identification of heterochromatic regions in the chromosomes of rye. Hereditas 74:233–238

7. Giraldez R, Cermeno MC, Orellana J (1979) Comparison of C-banding pattern in the chromosomes of inbred lines and open pollinated varieties of rye. Z Pflanzenzuchtg 83:40–48

8. Jellen EN, Phillips RL, Rines HW (1993) C-banded karyotypes and polymorphisms in hexaploid oat accessions (*Avena* spp.) using Wright's stain. Genome 36:1129–1137

9. Kato A (1999) Air drying method using nitrous oxide for chromosome counting in maize. Biotech Histochem 74:160–166

# Chapter 2

# Chromosome Painting by GISH and Multicolor FISH

**Steven S. Xu, Zhao Liu, Qijun Zhang, Zhixia Niu, Chao-Chien Jan, and Xiwen Cai**

## Abstract

Fluorescent in situ hybridization (FISH) is a powerful cytogenetic technique for identifying chromosomes and mapping specific genes and DNA sequences on individual chromosomes. Genomic in situ hybridization (GISH) and multicolor FISH (mc-FISH) represent two special types of FISH techniques. Both GISH and mc-FISH experiments have general steps and features of FISH, including chromosome preparation, probe labeling, blocking DNA preparation, target-probe DNA hybridization, post-hybridization washes, and hybridization signal detection. Specifically, GISH uses total genomic DNA from two species as probe and blocking DNA, respectively, and it can differentiate chromosomes from different genomes. The mc-FISH takes advantage of simultaneous hybridization of several DNA probes labeled by different fluorochromes to different targets on the same chromosome sample. Hybridization signals from different probes are detected using different fluorescence filter sets. Multicolor FISH can provide more structural details for target chromosomes than single-color FISH. In this chapter, we present the general experimental procedures for these two techniques with specific details in the critical steps we have modified in our laboratories.

**Key words** Fluorescent in situ hybridization, Genomic in situ hybridization, Molecular cytogenetics, DNA probe, Blocking DNA

## 1  Introduction

Fluorescent in situ hybridization (FISH) is a DNA-based cytogenetic technique developed in the early 1980s [1]. It has become a routine molecular cytogenetic approach for chromosome identification and physical mapping of specific genes and DNA sequences to individual chromosomes [2, 3]. The FISH technique utilizes DNA denaturation and renaturation for the formation of the hybrids between hapten (e.g., biotin and digoxigenin)-labeled probe DNA and the target DNA fixed on

---

Mention of trade names or commercial products in this article is solely for the purpose of providing specific information and does not imply recommendation or endorsement by the US Department of Agriculture.

Shahryar F. Kianian and Penny M.A. Kianian (eds.), *Plant Cytogenetics: Methods and Protocols*,
Methods in Molecular Biology, vol. 1429, DOI 10.1007/978-1-4939-3622-9_2,
© Springer Science+Business Media New York 2016

the slide and uses fluorochromes for hybridization signal detection [4, 5]. A FISH experiment includes a series of steps for chromosome preparation, probe labeling, blocking DNA preparation, probe target and DNA denaturation and hybridization, post-hybridization washes, and hybridization signal detection. Based on the type and number of DNA probes and the type of target DNA, several different types of FISH procedures, such as genomic in situ hybridization (GISH), low- or single-copy FISH, bacterial artificial chromosome (BAC) clone FISH (BAC-FISH), cDNA FISH, multicolor FISH (mc-FISH), and fiber-FISH, have been established for specific research purposes. This chapter elucidates only the procedures for GISH and mc-FISH. Both GISH and mc-FISH experiments have general operational steps and features of FISH. Specifically, GISH utilizes total genomic DNA from two species with different genomes as probe and blocking DNA, respectively, with the latter being used in a much higher quantity [6]. GISH is especially useful to differentiate chromosomes from different genomes [6], and it is extensively used for determining chromosome constitutions of amphiploids, identifying alien chromosomes in chromosome addition and substitution lines, and determining the size and location of the alien chromosome segments in translocation lines [5, 7–10]. In mc-FISH, at least two different DNA probes labeled with different fluorochromes are simultaneously used in one experiment. Hybridization signals from different probes are distinguished and captured through the mechanical rotation of fluorescence excitation filters [11]. The captured images are then pseudo-colored and merged using an imaging system [11]. Mc-FISH can provide much more information for target chromosomes than single-color FISH and is extensively used for precise chromosome identification and physical localization of genes and DNA sequences in major crop species such as wheat (*Triticum aestivum* L.) [12], rice (*Oryza sativa* L.) [13], canola (*Brassica napus* L.) [2], and many others.

The GISH and mc-FISH protocols described in this chapter were originally acquired from the Wheat Genetics Resource Center (WGRC) (Kansas State University, Manhattan, KS, USA) (see details in Zhang and Friebe [6] and WGRC Electronic Laboratory Manual at http://www.k-state.edu/wgrc/Protocols/labbook. html, verified 28 Jan. 2015). The protocols are now fully adapted to our laboratories and are routinely used for chromosome identification and analysis in wheat, sunflower (*Helianthus annuus* L.), and their related species in our alien gene introgression programs [14–18] (Fig. 1). In this chapter, the general operational steps described by Zhang and Friebe [6] with improvements and modifications by our group and specific details of the method are presented.

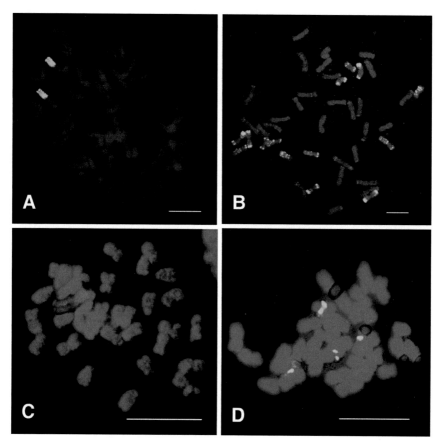

**Fig. 1** Chromosome painting images by genomic in situ hybridization (GISH) and multicolor fluorescence in situ hybridization (mc-FISH) in wheat and sunflower. (**a**) A GISH image of mitotic chromosomes in a wheat-wheat-grass (*Thinopyrum intermedium* (Host) Barkworth & D. R. Dewey) chromosome translocation line. The total genomic DNA isolated from *Th. intermedium* and common wheat variety "Chinese Spring" were used as probe labeled with biotin-16-dUTP and blocking DNA, respectively. The *Th. intermedium* chromatin (*fluoresced yellow green*) was detected with fluorescein isothiocyanate-conjugated avidin (FITC-avidin), and the wheat chromatin (*fluoresced red*) was counterstained with propidium iodide (PI) contained in VECTASHIELD Mounting Medium. (**b**) Multicolor FISH on somatic chromosomes of Chinese Spring probed with pAs1 (*green*) and pSc119.2 (*red*). The probe pAs1 is a clone containing a 1-kb repetitive DNA sequence from goat grass species *Aegilops tauschii* Cosson [19], and pSc119.2 is a clone containing the 120-bp repeat unit of a tandemly arranged DNA family derived from rye (*Secale cereale* L.) [20]. The probe pAs1 labeled with digoxigenin-11-dUTP for targeting D-genome chromosomes (*green*) and pSc119.2 labeled with biotin-16-dUTP for B-genome chromosomes (*red*) simultaneously. Chromosomes were counterstained with 4′,6-diamidino-2-phenylindole (DAPI) (*blue*) in VECTASHIELD Mounting Medium. (**c**) GISH analysis on mitotic chromosomes of a cultivated sunflower (*Helianthus annuus* L.) plant containing an alien chromosome (*red*) from wild species swamp sunflower (*H. angustifolius* L.). The genomic DNA of *H. angustifolius* was labeled with digoxigenin-11-dUTP and detected by anti-digoxigenin-rhodamine (*red*); the cultivated sunflower chromosomes were counterstained by DAPI (*blue*), with the ratio of blocking DNA to probe DNA of 30:1. (**d**) Multicolor FISH analysis on an F₁ hybrid of cultivated sunflower (HA 89) with wild species Nuttall's sunflower (*H. nuttallii* Torr. & A. Gray), probed with 45S rDNA (*red*) and 5S rDNA (*green*). 45S rDNA was labeled with digoxigenin-11-dUTP and detected by anti-digoxigenin-rhodamine, and 5S rDNA was labeled with biotin-16-dUTP and detected by FITC-avidin. Chromosomes were counterstained with DAPI (*blue*) in VECTASHIELD Mounting Medium. Bar = 10 μm

## 2  Materials

### 2.1  Equipment and Supplies

1. Epifluorescence microscope with CCD (charge-coupled device) camera and appropriate filters for fluorescence detection.

2. Computer system with image capturing and processing software.

3. Light microscope with phase contrast.

4. Microscope slide humidity chamber.

5. Parafilm coverslips ($25 \times 25$ mm) cut from parafilm.

6. Coverslips: No. 1 $18 \times 18$ mm, No. 1 $22 \times 22$ mm, No. 1.5 $22 \times 22$ mm, and No. 1 $22 \times 30$ mm (*see* **Note 1**).

### 2.2  Chemicals

1. Probe labeling kit: Nick translation DNA labeling kit.

2. Digoxigenin-11-dUTP: Alkali-stable, tetralithium salt, 1 mM.

3. Biotin-16-dUTP: Tetralithium salt, 1 mM.

4. Fluoresceins and mounting medium: 4′, 6-Diamidino-2-phenylindole (DAPI), fluorescein streptavidin (FITC-avidin), anti-digoxigenin-rhodamine, propidium iodide (PI), and VECTASHIELD Mounting Medium.

5. Single-stranded DNA (ssDNA) from salmon testes (10 mg/mL).

6. RNase A (20 mg/mL).

7. DNA isolation and purification kits.

8. Cellulase.

9. Pectinase.

10. Rubber cement.

### 2.3  Solutions

1. 20× SSC (saline sodium citrate) stock solution: To make 1 L of the solution, dissolve 175.3 g of sodium chloride (NaCl) and 88.2 g of trisodium citrate dihydrate ($C_6H_5Na_3O_7 \cdot 2H_2O$) in 600 mL of deionized distilled (dd$H_2O$) in a 1-L beaker and then bring final volume to 1 L with dd$H_2O$ in a 1-L volumetric flask. Adjust the pH to 7.0 using 0.1 M HCl, autoclave, and store at 4 °C in a refrigerator or at room temperature.

2. 4× SSC/0.2 % Tween-20: Dilute 100 mL of 20× SSC with 400 mL of dd$H_2O$ to make up 500 mL solution, add 1 mL of Tween-20. Store at room temperature.

3. 2× SSC: Dilute ten times of 20× SSC stock solution, e.g., dilute 100 mL of 20× SSC with 900 mL of dd$H_2O$ to make up 1 L of 2× SSC solution (*see* **Note 2**). Store at room temperature.

4. 0.1× SSC: Dilute 50 mL of 2× SSC with 950 mL of ddH$_2$O to make up 1 L of 0.1× SSC solution (*see* **Note 2**). Store at room temperature.

5. 1% acetocarmine: To make 500 mL of this stain, pre-heat 500 mL 45% acetic acid to boiling in a 1-L beaker under a fume hood. Remove from heat and slowly add 5 g carmine powder, and boil for 5–10 min. After cooling to room temperature, filter the stain solution into a 500-mL brown-colored glass bottle and store at 4 °C in a refrigerator.

6. 70% formamide: For 50 mL of this solution, mix 35 mL of formamide with 15 mL of 2× SSC (*see* **Note 3**). Store at room temperature.

7. 30% formamide: Mix 15 mL of formamide with 35 mL of 2× SSC (*see* **Note 4**). Store at room temperature.

8. 50% dextran sulfate sodium salt: Dissolve 10 g of dextran sulfate sodium salt in ddH$_2$O (*see* **Note 5**) to make a 20 mL solution and store in 1- or 0.5-mL aliquots in –20 °C.

9. TE buffer: To make 1 L TE buffer, mix 10 mL 1 M Tris–HCl (pH = 8.0), 2 mL 0.5 M Na$_2$EDTA (pH = 8.0), and 988 mL ddH$_2$O. Autoclave the solution for 15 min at 121 °C. Store at room temperature.

10. 2× CTAB DNA extraction buffer: 2% (w/v) CTAB, 100 mM Tris–HCl (pH = 8.0), 20 mM Na$_2$EDTA, and 1.4 M NaCl. Store at room temperature.

11. 5% BSA in 4× SSC/0.2% Tween-20: Dissolve 2.5 g BSA in 4× SSC/0.2% Tween-20, adjust with 4× SSC/0.2% Tween-20 to the final volume of 50 mL, and store in 1-mL aliquots in –20 °C.

12. Sodium citrate buffer: 4 mM citric acid and 6 mM sodium citrate.

13. Enzyme mixture of 2% cellulase and 24.3% pectinase: Dissolve 0.2 g cellulose in 7.40 mL of sodium citrate buffer, add 2.43 mL of pectinase for the final volume of 10 mL, and store in 1-mL aliquots in –20 °C.

14. 3 M NaAc (pH = 4.6).

# 3   Methods

## 3.1   GISH Procedure

### 3.1.1   Isolation of Genomic DNA for Probe and Blocking

1. Collect about 5 g of leaf tissue from seedling plants and grind them with liquid nitrogen into a fine powder.

2. Put leaf tissue powder in a 50-mL centrifuge tube, add 20 mL of 2× CTAB DNA extraction buffer, mix well, and incubate in water bath at 65 °C for 30–60 min.

3. Add 10 mL of phenol and 10 mL of chloroform/isoamyl alcohol (24:1), mix well, and centrifuge at 4000 rpm ($3327 \times g$) for 20 min.

4. Transfer 15 mL of supernatant using plastic transfer pipettes into a new 50-mL tube and precipitate DNA by adding two volumes of 95 % (or 100 %) ethanol or one volume of isopropyl alcohol.

5. Wash DNA pellet using 70 % ethanol twice and put the DNA pellet into a 2-mL microcentrifuge tube to air dry.

6. Add 900 μL 1× TE buffer to dissolve the DNA pellet, then add 5–10 μL RNase A (10 mg/mL), mix well, and incubate at 37 °C for 1 h.

7. Add 900 μL chloroform/isoamyl alcohol (24:1), mix well, and then centrifuge at 13,000 rpm ($13{,}793 \times g$) for 20 min in a microcentrifuge.

8. Transfer the supernatant using a small pipette into a new 2-mL centrifuge tube, add an equal volume of chloroform/isoamyl alcohol (24:1), mix well, and then centrifuge at 13,000 rpm ($13{,}793 \times g$) for 20 min in a microcentrifuge.

9. Transfer the supernatant into a new 2-mL centrifuge tube using a small pipette, then precipitate DNA by adding two volumes of 95 or 100 % ethanol.

10. Centrifuge at 10,000 rpm ($8161 \times g$) for 5 min in a microcentrifuge.

11. Wash the DNA pellet twice using 70 % ethanol, and air dry.

12. Dissolve DNA pellet in 500 μL 1× TE buffer.

*3.1.2 Probe Labeling*

1. DNA is labeled with biotin-16-dUTP using a Nick Translation Kit following the manufacturer's instruction.

2. Check the size of the probes by separating the nick translation product on an agarose gel with a DNA size marker. The size of probes should be in the range of 300–600 bp.

*3.1.3 Blocking DNA Preparation*

1. Adjust the concentration of blocking DNA isolated in Subheading 3.1.1 to between 0.1 and 1.0 μg/μL.

2. Add 10 M NaOH to the DNA in a 2-mL centrifuge tube to bring the final concentration of NaOH to 0.4 M.

3. Put the centrifuge tube containing the DNA sample in boiling water for 40–50 min.

4. Place the DNA sample on ice for 5 min (*see* **Note 6**).

5. Add an equal volume of 3 M NaAc (pH = 4.6) and two volumes of 95 or 100 % cold ethanol (i.e., stored in –20 °C). Mix well to precipitate the DNA.

6. Centrifuge at 10,000 rpm ($8161 \times g$) for 20 min in a microcentrifuge.

7. Wash the DNA pellet using 70 % ethanol and air dry.

8. Dissolve the pellet with 400 µL 1× TE buffer and then add 1/10 volume of 3 M NaAc (pH = 7.0) and two volumes of 95 or 100 % cold ethanol. Mix well and then centrifuge at 10,000 rpm (8161 × $g$) for 5 min in a microcentrifuge. Wash pellet twice using 70 % ethanol and air dry.

9. Add 100 µL 1× TE buffer to dissolve the DNA pellet and then adjust DNA concentration to 2–10 µg/µL.

10. Check DNA size using 1 % agarose gel. The size of DNA should be in the range of 200–400 bp.

*3.1.4  Chromosome Preparation*

1. Germinate seed on moist blotting paper in Petri dishes at room temperature or in an incubator at 22 °C (*see* **Note 7**).

2. Collect roots when they are 1–2 cm long. Place root tips in a 1.5-mL centrifuge tube with cold ddH$_2$O and put the centrifuge tube on ice in a cooler for 18–30 h (*see* **Note 8**).

3. Fix root tips in a 3:1 mixture of 95 % ethanol and glacial acetic acid at room temperature for 12–24 h (*see* **Note 9**).

4. Stain root tips in 1 % acetocarmine for 10–20 min. Remove the root cap and excise a small piece (0.5–1.0 mm long) of meristematic region of the root tip (darker stained region of the root) with a stainless razor blade, and then place the meristematic tissue on a pretreated clean slide.

5. Add a drop of 45 % acetic acid on the tissue and cover it with a glass coverslip. Mitotic cells in the tissue are separated and spread by gently tapping the coverslip with a dissecting needle having a blunt end point.

6. Briefly heat the slide on the flame of an alcohol burner, place a piece of filter paper on the coverslip, and then vertically press the coverslip with a finger to squash the root cells (*see* **Note 10**).

7. Observe the slides under a phase contrast microscope at 100× magnification and select the slides with good chromosome preparation (i.e., the slides should have adequate metaphase cells in which chromosomes are well spread and in good shape) for the next step.

*3.1.5  Slide Pretreatment*

1. Firmly hold the slide using a 6-in. forceps, dip the slide vertically into liquid nitrogen in a 600-mL Dewar flask for 10 s, take the slide out of the liquid nitrogen, and immediately flip off coverslip with a razor blade by lifting the corner of coverslip.

2. Dry the slides at 60 °C in a lab oven for 1–3 h or directly dehydrate the slides using a series of 70 %, 95 %, and 100 % ethanol washes at room temperature for 3–5 min each, respectively.

3. Add 70–100 µL of RNase A (100 µg/mL in 2× SSC) to the slide, cover with precut parafilm coverslip, and incubate the slides in a humid chamber at 37 °C for 30 min to 1 h (*see* **Note 11**).

4. Remove parafilm coverslip and wash slides three times, 5 min each, using 2× SSC at room temperature.

5. Denature the chromosomal DNA by putting slides in a 70% formamide (in 2× SSC) solution, in a container, in a water bath at 75 °C for 2 min (*see* **Note 12**).

6. Dehydrate slides in a series of 70%, 95%, and 100% ethanol at −20 °C for 3–5 min each, respectively (*see* **Note 13**).

7. Dry slides at room temperature.

*3.1.6 Hybridization*

1. Prepare hybridization mixture using the following recipe (Table 1) based on the amount used for one slide (*see* **Note 14**).

2. Mix well and spin down hybridization mixture up to 5000 rpm ($2040 \times g$) in a microcentrifuge. Denature hybridization mixture at 100 °C (boiling the mixture) for 10 min and put it on ice immediately for 10 min.

3. Mix well and spin down denatured hybridization mixture up to 5000 rpm ($2040 \times g$) in a microcentrifuge. Keep it on ice.

4. Add about 20 μL of denatured hybridization mixture to the slides, cover with coverslip ($22 \times 22$ mm), and seal coverslip using rubber cement.

5. Put the slides in humidity chamber, created by placing slides on two glass stir rods above the water-soaked filter paper in a rectangle plastic box with cover (*see* **Note 15**).

6. Put the covered plastic box into an incubator at 37 °C for overnight hybridization.

**Table 1**
**Recipe of hybridization for chromosome labeling for one microscope slide**

| Solutions | Amount (μL) |
| --- | --- |
| 100% deionized formamide | 10 |
| 50% dextran sulfate | 4 |
| 20× SSC (pH = 7.0) | 2 |
| 10 mg/mL sperm ssDNA | 1 |
| Probe DNA | 1 |
| Blocking DNA[a] | 0–2 |
| ddH$_2$O[b] | 0–2 |
| Total | 20 |

[a]The amount of the blocking DNA depends on the concentration of the blocking DNA sample and ratio of probe DNA to blocking DNA
[b]The amount of ddH$_2$O depends on the amount of the blocking DNA

*3.1.7 Post-hybridization Wash and Signal Detection*

1. Remove rubber cement using tweezers and put slides in 2× SSC at 42 °C until coverslip falls off.

2. Wash slides in 2× SSC twice for 5 min each at 42 °C.

3. Drain slides and put them in 2× SSC at room temperature for 5 min (*see* **Note 16**).

4. Drain slides and put them in 4× SSC with 0.2 % Tween-20 at room temperature for 5 min.

5. Drain slides and add 80–100 μL 5 % BSA in 4× SSC/0.2 % Tween-20 on each slide. Cover the slides with a parafilm coverslip and keep the slides in humid chamber at 37 °C for about 30 min (*see* **Note 16**).

6. Drain slides and apply 80–100 μL FITC-avidin (FITC-avidin: 5 % BSA = 1 : 200 dilution) to the slides and incubate the slides in the plastic box in the dark at 37 °C for 1 h.

7. Drain and wash the slides twice in 4× SSC/0.2 % Tween-20 at 42 °C for 5 min each.

8. Wash slides in distilled water twice for 2 min each to remove salt on the slides; dry the slide in room temperature until visibly dry.

9. Add 13 μL anti-fade mounting medium containing 1 μg/mL PI or 2 μg/mL DAPI for counterstaining on the target area and cover the area with 22 × 22 mm glass coverslip.

10. Observe slide right away or wait overnight for fluorescence to stabilize. Store the slides in the dark (*see* **Note 17**).

11. Observe images under a fluorescence microscope; capture images using a CCD camera.

### 3.2 Multicolor FISH Procedure

*3.2.1 Isolation of Genomic DNA for Blocking and the DNA for Probe*

1. Genomic DNA used for blocking is extracted using the same protocol as that for GISH (Subheading 3.1.1) (*see* **Note 18**).

2. The DNA for probe can be total genomic DNA, cloned DNA fragments inserted into a plasmid vector, or that obtained by PCR. The plasmid DNA for probe is extracted from the *E. coli* using a commercial extraction kit. DNA fragment obtained by PCR is to be purified using a commercial purification kit.

*3.2.2 Probe Labeling*

For mc-FISH, more than one probe is used for the hybridization.

1. DNA is sheared and denatured in boiling water bath or a heating block at 100 °C for 10 min and cooled down on ice for at least 5 min.

2. DNA is labeled with biotin-16-dUTP or digoxigenin-11-dUTP using a Nick Translation Kit following the manufacturer's instruction.

3. Check the size of the probes by running the Nick Translation product on an agarose gel as for GISH (Subheading 3.1.2, **step 2**).

*3.2.3 Blocking DNA Preparation*

Blocking DNA is prepared as for GISH (Subheading 3.1.3) (*see* **Note 19**).

*3.2.4 Slide Preparation*

1. Germinate seeds on moist blotting paper in Petri dishes on top of laboratory bench at room temperature or in an incubator at 22 °C (*see* **Note 20**).

2. Collect roots, 1–2 cm long, into a 1.5-mL centrifuge tubes with cold ddH$_2$O and put centrifuge tubes into the ice in a cooler for 18–34 h (*see* **Note 21**).

3. Fix the root tips in a 3:1 mixture of 95 % ethanol and glacial acetic acid at room temperature as for GISH (**step 3** of Subheading 3.1.4).

4. Root tip cells are squashed as for GISH (**steps 4** and **5** of Subheading 3.1.4).

5. Check the slides under a phase contrast microscope for good metaphase chromosome spreads.

*3.2.5 Slide Pretreatment*

Use the same pretreatment method as used for GISH (Subheading 3.1.5).

*3.2.6 Hybridization*

1. Prepare hybridization mixture solution using the following recipe (Table 2) based on the amount for one slide (*see* **Note 22**).

**Table 2**
**Recipe of hybridization mixture for colored chromosome labeling for one microscope slide**

| Solutions | Amount (µL) |
|---|---|
| 100 % deionized formamide | 10 |
| 50 % dextran sulfate | 4 |
| 20× SSC (pH = 7.0) | 2 |
| 10 mg/mL sperm ssDNA | 1 |
| Probe DNA 1-Digoxigenin[a] | 0.5–1 |
| Probe DNA 2-Biotin[a] | 0.5–1 |
| Blocking DNA[b] | 0–2 |
| ddH$_2$O[c] | 0–2 |
| Total | 20 |

[a]The amount of the probe depends on the probe DNA concentration and hybridization signal intensity
[b]For the probe DNA from repetitive sequences such as pAs1, pSc119.2, 45S rDNA, and 5S rDNA, centromeric sequences, etc., no blocking DNA is needed. For the probe DNA from low-copy sequences such as BAC clones, blocking DNA is still needed, and the amount depends on its concentration and ratio of probe DNA to blocking DNA
[c]The amount of ddH$_2$O depends on the amount of the probe and blocking DNA

2. The hybridization mixture was denatured at 100 °C as for GISH (**steps 2** and **3** of Subheading 3.1.6).

3. The hybridization mixture is put on slides and hybridized at 37 °C for overnight as for GISH (**steps 4** through **6** of Subheading 3.1.6).

*3.2.7 Post-hybridization Slide Wash and Signal Detection*

1. Remove rubber cement using tweezers, and put slides in 2× SSC at 42 °C until coverslip is falling off as for GISH (**step 2** of Subheading 3.1.7); incubate the slides in 2× SSC at 42 °C for another 5 min.

2. Wash the slides in 30% formamide in 2× SSC at 42 °C for 5 min.

3. Wash the slides in 0.1× SSC twice at 42 °C for 5 min each.

4. Put the slides in 2× SSC at 42 °C for 5 min.

5. Put slides in 2× SSC at room temperature for 5 min as for GISH (**step 3** of Subheading 3.1.7).

6. Drain slides and put them in 4× SSC/0.2% Tween-20 at room temperature for 5 min as for GISH (**step 4** of Subheading 3.1.7).

7. Drain slides and add 80–100 μL 5% BSA in 4× SSC/0.2% Tween-20 to each slide. Cover the slides with parafilm coverslip and keep the slides in humid chamber at 37 °C for around 30 min as for GISH (**step 5** of Subheading 3.1.7).

8. Drain slides and apply 80–100 μL of FITC-avidin (FITC-avidin: 5% BSA = 1: 100) and anti-digoxigenin-rhodamine (anti-digoxigenin-rhodamine: 5% BSA = 1: 100) to the slides and incubate them in a plastic box in the dark at 37 °C for 1 h.

9. Drain and wash the slides twice in 4× SSC/0.2% Tween-20 at 42 °C for 5–8 min each as for GISH (**step 7** of Subheading 3.1.7).

10. Wash slides in distilled water twice as for GISH (**step 8** of Subheading 3.1.7).

11. Add 13 μL anti-fade mounting medium containing 2 μg/mL DAPI for counterstaining on the target area and cover the area with 22 × 22 mm glass coverslip (*see* **Note 23**).

12. Observe the slides right away or wait overnight for fluorescence to stabilize as for GISH (**step 10** of Subheading 3.1.7). Store the slides in dark.

13. Observe images under a fluorescence microscope; capture images using a CCD camera (*see* **Note 24**).

# 4  Notes

1. No. 1 coverslips are used for slide preparation and hybridization, and No. 1.5 are used for image observation and capture.

2. 2× SSC and 0.1× SSC solutions can be autoclaved at 121 °C for 15 min for long-term storage at room temperature.

3. 70 % formamide should be frozen in –20 °C in 35-mL aliquots in 50-mL centrifuge tubes.

4. 30 % formamide should be frozen in –20 °C in 15-mL aliquot in 50-mL centrifuge tubes.

5. Dissolve dextran sulfate sodium using a water bath at 60 °C while stirring, as this chemical dissolves very slowly in $H_2O$ at room temperature.

6. Centrifuge the DNA solution at 10,000 rpm ($8161 \times g$) for 5 min in a microcentrifuge and transfer supernatant into a new 2-mL microtube if sedimentations appear in the bottom of microtube after the DNA solution cools down.

7. To achieve uniform germination, we usually allow the seeds to absorb $ddH_2O$ in the Petri dishes for 24 h at room temperature and then put the Petri dishes in a vernalization chamber or refrigerator (2–4 °C) for 24 h for old seeds (>12 months) and for 3–5 days for newly harvested seeds. Lack of excessive water on blotting paper is a key to getting healthy root tips with adequate cells that are actively dividing.

8. The cooler is usually put in a refrigated chamber (2–4 °C) to prevent the ice from thawing quickly.

9. The fixation time specified in the literatures is 48 h at room temperature. We didn't observe any difference in fixation for 12–24 h compared with 48 h. However, if root tips are in the fixative more than a week at room temperature, chromosome morphology may change. If GISH can't be performed in 1–2 weeks, the root tips should be stored in –20 °C or –80 °C freezer. Root tips can be stored in –20 °C or –80 °C freezer for several years.

10. An alternative method such as enzyme digestion and flame dry method could be used for slide preparation (*see* ref. 17, 18). For enzyme digestion, the root tips are digested at 37 °C for 1–2.5 h in an enzyme mixture of 1 or 2 % cellulase (w/v) and 24 % pectinase (v/v) in 10 mM sodium citrate buffer (4 mM citric acid and 6 mM sodium citrate). After enzyme digestion, the root tips are used for making chromosome spreads by either flame dry method or squashing root tip tissues in 45 % acetic acid as described in the **steps 4** and **5** of Subheading 3.1.4. For the flame dry method, the root tips are gently washed in distilled water after enzyme digestion and then fixed in a 3:1 mixture of methanol (or ethanol) and glacial acetic acid. The root tip tissues from 1 to 2 root tips are excised on a clean slide and macerated in a drop of fixation solution using a fine-pointed forceps. The slide is then quickly flamed-dried over an alcohol lamp (*see* ref. 18).

11. This step may not be necessary in cases where the probes are well labeled.

12. For the newly prepared slides (e.g., less than 24 h), the denaturing time should be about 1 min. Longer denaturing time can cause chromosome deformation.

13. The ethanol series should be replaced when the slides taken out of the 100% ethanol do not dry quickly at room temperature.

14. Because 50% dextran sulfate is sticky, we make up a large amount (e.g., 50 mL) of the mixture of 100% deionized formamide, 50% dextran sulfate, 20× SSC (pH = 7.0), and 10 mg/mL ssDNA based on the recipe and store the mixture in 1-mL aliquots in –20 °C freezer.

15. Add more water into the box if the slides are sealed with rubber cement, but the slides should not be submerged in the water. More water in the box makes it easy to remove the rubber cement after hybridization.

16. This step can be omitted.

17. Slides can be stored in 4 °C or –20 °C.

18. High DNA concentration (above 1 µg/µL) and quality are preferred since the volume of the hybridization mixture is limited and high efficiency of blocking is needed to reduce the hybridization background.

19. Alternative methods can be used for blocking DNA preparation, such as shearing the genomic DNA by putting it in boiling water bath for 20 min or autoclaving the genomic DNA at 121 °C for 5 min, followed by cooling down on ice for at least 5 min.

20. For species with taproots such as sunflower, the root tips were collected from 2- to 3-week-old seedlings.

21. The treatment time is adjusted according to the size of the chromosomes. The larger the chromosomes are, the longer the treatment time.

22. The probe DNA can be purified before making the hybridization mixture using the commercial purification kit such as QIAquick Nucleotide Removal Kit.

23. For the slides prepared by the flame-dried method, a bigger coverslip such as No. 1 22 × 30 mm will be used to cover the target area.

24. At least three fluorescence filters are needed for taking a merged image.

## Acknowledgment

The authors thank Drs. Bikram S. Gill and Bernd Friebe (Kansas State University, Manhattan, KS, USA) for providing the original protocols. We also thank Drs. Xueyong Zhang (Chinese Academy of Agricultural Sciences, Beijing, China), Qi Zheng (Chinese Academy of Sciences, Beijing, China), Lili Qi (USDA-ARS, Fargo, ND, USA), and Jiming Jiang (University of Wisconsin, Madison, WI, USA) for the instruction and discussion of the procedure. Dr. Jiming Jiang kindly provided the plasmids of 5S rDNA and 45S rDNA. This work was supported by the USDA-ARS CRIS Project No. 3060-520-037-00D. USDA is an equal opportunity provider and employer.

## References

1. Langer PR, Waldrop AA, Ward DC (1981) Enzymatic synthesis of biotin-labeled polynucleotides: novel nucleic acid affinity probes. Proc Natl Acad Sci U S A 78:6633–6637

2. Feng J, Primomo V, Li Z, Zhang Y, Jan C-C, Tulsieram L, Xu SS (2009) Physical localization and genetic mapping of the fertility restoration gene *Rfo* in canola (*Brassica napus* L.). Genome 52:401–407

3. Danilova TV, Friebe B, Gill BS (2014) Development of a wheat single gene FISH map for analyzing homoeologous relationship and chromosomal rearrangements within the Triticeae. Theor Appl Genet 127:715–730

4. Liehr T, Pellestor F (2009) Molecular cytogenetics: the standard FISH and PRINS procedure. In: Liehr T (ed) Springer protocols - fluorescence in situ hybridization (FISH) application guide. Springer, Berlin Heidelberg, pp 23–34

5. Jiang J, Gill BS (1994) Non-isotopic *in situ* hybridization and plant genome mapping: the first 10 years. Genome 37:717–725

6. Zhang P, Friebe B (2009) FISH on plant chromosomes. In: Liehr T (ed) Springer protocols - fluorescence in situ hybridization (FISH) application guide. Springer, Berlin Heidelberg, pp 365–394

7. Cai X, Jones SS, Murray TD (1998) Molecular cytogenetic characterization of *Thinopyrum* and wheat-*Thinopyrum* translocated chromosomes in a wheat-*Thinopyrum* amphiploid. Chromosome Res 6:183–189

8. Xu SS, Faris JD, Cai X, Klindworth DL (2005) Molecular cytogenetic characterization and seed storage protein analysis of 1A/1D translocation lines of durum wheat. Chromosome Res 13:559–568

9. Oliver RE, Xu SS, Stack RW, Friesen TL, Jin Y, Cai X (2006) Molecular cytogenetic characterization of four partial wheat-*Thinopyrum ponticum* amphiploids and their reactions to Fusarium head blight, tan spot, and Stagonospora nodorum blotch. Theor Appl Genet 112:1473–1479

10. McArthur RI, Zhu X, Oliver RE, Klindworth DL, Xu SS, Stack RW, Wang RR-C, Cai X (2012) Homoeology of *Thinopyrum junceum* and *Elymus rectisetus* chromosomes to wheat and disease resistance conferred by the *Thinopyrum* and *Elymus* chromosomes in wheat. Chromosome Res 20:699–715

11. Bayani J, Squire J (2004) Multi-color FISH techniques. Curr Protoc Cell Biol. 24:22.5.1–22.5.25

12. Zheng Q, Lv Z, Niu Z, Li B, Li H, Xu SS, Han F, Li Z-S (2014) Molecular cytogenetic characterization and stem rust resistance of five wheat-*Thinopyrum ponticum* partial amphiploids. J Genet Genomics 41:591–599

13. Shishido R, Sano Y, Fukui K (2000) Ribosomal DNAs: an exception to the conservation of gene order in rice genomes. Mol Gen Genet 263:586–591

14. Niu Z, Klindworth DL, Friesen TL, Chao S, Jin Y, Cai X, Xu SS (2011) Targeted introgression of a wheat stem rust resistance gene by DNA marker-assisted chromosome engineering. Genetics 187:1011–1021

15. Niu Z, Klindworth DL, Yu G, Friesen TL, Chao S, Jin Y, Cai X, Ohm J-B, Rasmussen JB, Xu SS (2014) Development and characterization of wheat lines carrying stem rust resistance gene *Sr43* derived from *Thinopyrum ponticum*. Theor Appl Genet 127:969–980

16. Klindworth DL, Niu Z, Chao S, Friesen TL, Jin Y, Faris JD, Cai X, Xu SS (2012) Introgression and characterization of a goatgrass gene for a high level of resistance to Ug99 stem rust in tetraploid wheat. G3 2:665–673

17. Liu Z, Wang D, Feng J, Seiler GJ, Cai X, Jan CC (2013) Diversifying sunflower germplasm by integration and mapping of a novel male fertility restoration gene. Genetics 193: 727–737

18. Feng J, Liu Z, Cai X, Jan CC (2013) Toward a molecular cytogenetic map for cultivated sunflower (*Helianthus annuus* L.) by landed BAC/BIBAC clones. G3 3:31–40

19. Rayburn AL, Gill BS (1986) Molecular identification of the D-genome chromosomes of wheat. J Hered 77:253–255

20. McIntyre CL, Pereira S, Moran LB, Appels R (1990) New *Secale cereale* (rye) DNA derivatives for the detection of rye chromosome segments in wheat. Genome 33:635–640

# Chapter 3

# Fluorescent In Situ Hybridization on Extended Chromatin Fibers for High-Resolution Analysis of Plant Chromosomes

## Daryna Dechyeva and Thomas Schmidt

## Abstract

Fiber FISH is a high-resolution cytogenetic method and a powerful tool of genome analysis to study the localization and the physical organization of markers, genes, and repetitive sequences on a molecular level. Measurement of physical distances between sequences can be performed along extended chromatin fibers with the resolution of up to 1 kb and is applicable to all plant species.

**Key words** Fiber FISH, Chromatin fibers, High-resolution chromosome analysis, Molecular cytogenetics, Repetitive DNA

## 1 Introduction

We describe fluorescent in situ hybridization (FISH) on extended chromatin fibers (fiber FISH) as a powerful method for visualization of DNA sequences on plant chromosomes using a UV microscope. Probe DNA is labeled with haptens, such as biotin or digoxigenin, or directly with fluorochromes. The procedure consists of the preparation of stretched chromatin fibers, hybridization, post-hybridization washes, and either the immunological detection of the probes with anti-hapten antibodies conjugated to fluorescent dyes or direct observation of fluorescently labeled probes.

FISH is able to reveal the exact physical organization of DNA along chromosomes. It allows the detection and precise localization of repetitive or single-copy DNA sequences on interphase nuclei, chromosomes, or chromatin fibers. Any cloned sequence, PCR product, synthetic oligonucleotide, BAC, as well as total genomic DNA can be used as probe. Initially, in situ hybridization with a radioactively labeled probe was developed to visualize RNA

Shahryar F. Kianian and Penny M.A. Kianian (eds.), *Plant Cytogenetics: Methods and Protocols*,
Methods in Molecular Biology, vol. 1429, DOI 10.1007/978-1-4939-3622-9_3,
© Springer Science+Business Media New York 2016

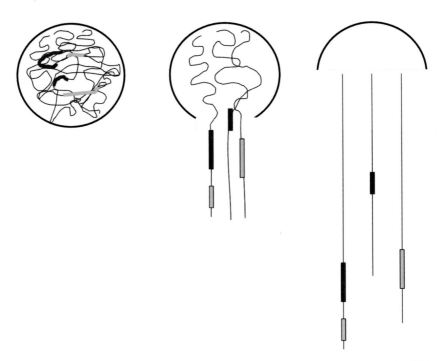

**Fig. 1** Principle of fiber FISH. Intact and lysed nuclei are depicted as *closed* and *open circles*. DNA targets are shown as *colored bars* (adapted from [16])

and DNA in mammalian cells [1]. FISH was established for mouse satellite DNA [2], followed by the first application in plants [3], and since then has been applied in molecular cytogenetics for the physical mapping of repeats, genes, and markers, karyotyping, and analysis of the genome architecture [4, 5]. Multicolor FISH has become a powerful tool for studies of genome composition and evolution [6–9, 17].

Somatic metaphase chromosomes provide an average resolution of 1 Mbp. In contrast, extended chromatin fibers extracted from somatic cells enable the highest cytogenetic resolution in physical mapping of DNA sequences in distances of only 1 kb [10]. Taking into consideration the stretching degree of chromatin fibers of 3.27 kb/μm, the multicolor fiber FISH can be used to measure the physical distance between the DNA sequences (Fig. 1). It bridges the resolution of megabase molecular techniques, such as pulsed-field gel electrophoresis and genome sequence analysis [11]. Fiber FISH was successfully used for the detailed investigation of chromosomal domains in *Arabidopsis* [10], rice [12], tomato [13], and sugar beet [14]. In repetitive regions of plant chromosomes which cannot be assembled in genome sequencing projects, fiber FISH allows positioning and length measurement of tandem arrays [15].

## 2  Materials

Prepare all solutions using sterile distilled water. Store all reagents according to manufacturers instructions or frozen at –20 °C, if not indicated otherwise

1. Plants are grown under greenhouse conditions. Collect very young leaves of vigorously growing plants and use immediately without pretreatment. Alternatively, germinate seeds at 25 °C in the dark and use whole seedlings (*see* **Note 1**).

2. Fixative: 3 v/v methanol (100 %) or ethanol to 1 v/v acetic acid (100 %).

3. Citrate buffer: 4 mM citric acid, 6 mM sodium citrate, pH 4.5 adjusted with HCl.

4. Primers for PCR of sequences cloned in multiple cloning sites of plasmid vectors (e.g., M13 forward, M13 reverse).

5. 20× SSC: 3 M NaCl, 0.3 M sodium citrate, pH 7.0.

6. Formamide: sterile filtered, aliquots stored frozen at –20 °C.

7. Dextran sulfate: 10 % (w/v) in water.

8. Sodium dodecyl sulfate: 0.1 % (w/v) in water, prepared under a fume hood.

9. Sonicated salmon sperm DNA: 250 ng/μL in 1× TE buffer.

10. Hybridization solution: 50 % formamide, 20 % dextran sulfate, 0.2 % SDS, 50 ng/μL sonicated salmon sperm DNA, 10–100 ng/μL labeled probes, in 2× SSC.

11. 1× TE buffer: 10 mM Tris–HCl, 1 mM EDTA, pH 8.0 adjusted with NaOH.

12. 1× TAE buffer: 40 mM Tris-acetate, 1 mM EDTA, pH 8.0 adjusted with NaOH.

13. Nuclei isolation buffer (NIB): 10 mM Tris–HCl, 10 mM EDTA, 100 mM KCl, 500 mM sucrose, 1 mM spermine, 4 mM spermidine, 0.1 % v/v β-mercaptoethanol, pH 9.5 adjusted with NaOH.

14. Lysis buffer (STE): 0.5 % w/v SDS, 100 mM Tris–HCl, 5 mM EDTA, pH 7.0 adjusted with NaOH.

15. McIlvaine's buffer: 40 mM disodium phosphate, 80 mM citric acid, pH 5.0 adjusted with HCl.

16. 4′,6-Diamidino-2-phenylindole dihydrochloride (DAPI) solution: 2 μg/mL in McIlvaine's buffer.

17. Anti-DIG-alkaline phosphatase or anti-biotin-AP.

18. Nitro blue tetrazolium—5-bromo-4-chloro-3-indolyl-phosphate (NBT/BCIP): NBT 100 mg/mL and BCIP 50 mg/mL (e.g., obtained from Roche).

19. 3 % w/v Bovine serum albumin (BSA).

20. Liquid protein block (5 %): 5 % w/v bovine serum albumin.

21. Control digoxigenin or biotin-labeled DNA of known concentration.

22. Tween 20.

23. Streptavidin or anti-digoxigenin.

24. TN: 0.1 M Tris–HCl, 0.15 M NaCl, pH 7.5.

25. TNM: 0.1 M Tris–HCl, 0.01 M NaCl, 0.05 M MgCl$_2$, pH 9.5.

**2.1 Supplies and Equipment**

1. 100, 50, and 20 μm nylon meshes.

2. UV microscope with filters appropriate for the hapten labels.

3. Nick translation kit (e.g., DIG-nick translation or biotin-nick translation kits) for FISH probes greater than 3 kb.

4. Positively charged nylon membrane (e.g., Hybond N$^+$ membrane).

5. UV transilluminator.

6. In situ thermocycler.

# 3  Methods

**3.1 Isolation of Plant Nuclei**

1. Chop 10–20 whole seedlings or young leaves in a glass Petri dish on ice in 2–5 mL NIB with a razor blade until a suspension is formed.

2. Filter the suspension consecutively through 100, 50, and 20 μm nylon meshes and centrifuge the suspension in a microcentrifuge for 4 min at 1150 g, 4 °C.

3. Discard the supernatant.

4. Carefully dissolve the nuclei pellet in 20 μL NIB.

5. To control the quality of the preparation (*see* **Note 2**), mix 2 μL of the nuclei suspension with DAPI solution on a glass slide and examine under the UV microscope (Fig. 2).

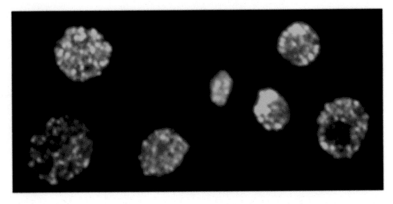

**Fig. 2** Suspension of DAPI stained plant nuclei under UV-microscope

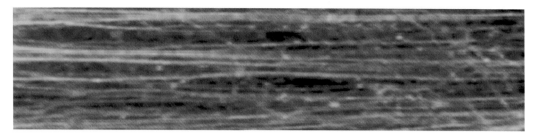

**Fig. 3** Extended chromatin fibers stained with DAPI (*blue*) under UV microscope

**3.2 Preparation of Extended DNA Fibers**

1. Spread 1.5 μL of the nuclei suspension on the edge of a glass slide and dry at room temperature (RT).

2. Apply 40 μL of STE onto each end of the slide, cover with a 50 mm long glass cover slip, and incubate horizontally for 1 min.

3. Tilt the slide carefully until the cover slip slides off slowly (*see* **Note 3**).

4. Air-dry the preparation in a rack, fix in fresh 1:3 (V/V) mixture of 100 % acetic acid : methanol or ethanol for 3 min at RT and incubate for 30 min at 60 °C on a hot plate. Use slides for FISH immediately afterwards (Fig. 3).

**3.3 Labeling of DNA Probes for FISH**

In order to detect specific DNA sequences on plant chromatin fibers, the DNA probes were labeled either directly with fluorochromes, biotin, or digoxigenin.

*3.3.1 Labeling of DNA Probes for FISH by PCR*

Labeling of FISH probes by PCR is suitable for templates consisting of cloned plasmids, PCR products, or genomic DNA (less than 3 kb long).

1. Labeling PCR reaction: Use 20–50 ng template DNA, 20 pM forward primer, 20 pM reverse primer, 10× PCR buffer, 10 mM dNTPs, 1.75 nM digoxigenin-11-dUTP or 3.5 nM biotin-16-dUTP, or 3.5 nM fluorochrome-dUTP, 2.5 U *Taq* DNA polymerase, in a total volume of 50 μL.

2. PCR program conditions: Pre-denaturation at 94 °C for 3 min, denaturation at 94 °C for 30 s, amplification at Tm°C for 30 s (*see* **Note 4**), elongation at 72 °C for 90 s; repeat denaturation/amplification/elongation for 35 cycles. Finish with final elongation at 72 °C for 5 min.

3. Quality check: Separate the DNA fragments in a horizontal 1.2 % agarose gel electrophoresis system at 3–9 V/cm in 1× TAE buffer. For visualization of the DNA, add ethidium bromide into the gels to a final concentration of 0.5 μg/mL.

The labeled probe containing the hapten migrates slower than the unlabeled control PCR product and is visible in the gel as a shifted band.

4. Purify the labeled probes from the unincorporated label by ethanol precipitation (*see* **Note 5**).

*3.3.2 Labeling of DNA Probes for FISH by Nick Translation*

Labeling by nick translation is suitable for DNA probes larger than 3 kb. The nick translation method is based on the ability of the DNase I to introduce randomly distributed single-strand breaks, or nicks, into double-stranded DNA. The nicks are repaired by DNA polymerase I, which removes the nucleotides and replaces them with digoxigenin- or biotin-labeled nucleotides.

Perform the labeling by nick translation following the manufacturer's instructions for kits. For single-color FISH, the labeling with digoxigenin is recommended. Purify the labeled probes from the unincorporated label by ethanol precipitation (*see* **Note 5**).

*3.3.3 Assessment of the Labeling Quality*

Assess the labeling quality for every batch of probe DNA.

When direct labeling, a nanospectrophotometrical measurement of the color intensity is performed according to the manufacturer's instructions.

In the case of digoxigenin or biotin labeling, estimate the quality by the color reaction.

1. Spot 0.5 and 1.0 μL of the labeled probe and 0.5 μL of the control labeled DNA onto positively charged nylon membrane and dry for 5 min at RT.

2. Place the membrane onto a UV transilluminator for 30 s and equilibrate in TN solution for 1 min.

3. Incubate the membrane with 5 % liquid protein block in TN solution for 30 min.

4. Incubate the membrane with the antibody solution containing 1 μL of anti-DIG-alkaline phosphatase or 5 μL of anti-biotin-AP depending on the labeling in 5 mL of TN solution for 30 min at 37 °C.

5. Wash the membrane in TN solution for 15 min and then in TNM solution for 2 min.

6. Pour the detection solution containing 75 μL of NBT/BCIP (nitro blue tetrazolium—5-bromo-4-chloro-3-indolyl-phosphate, in 5 mL of TNM solution over the membrane and incubate for 10 min in the dark (*see* **Note 6**).

7. The intensity of the resulting color of the labeled probe with the control allows estimation of the quality of the labeling.

**3.4 Fluorescent In Situ Hybridization on DNA Fibers**

*3.4.1 Hybridization of the Probe*

1. Apply 30 µL of the hybridization solution in aliquots onto dried slides, and cover the slides with glass or plastic cover slips (cut from, e.g., colorless polyethylene autoclavable plastic bags).

2. Denature and stepwise cool down in an in situ thermocycler with filled water reservoir. Touchdown denaturation program is 70 °C 8 min, 55 °C 5 min, 50 °C 2 min, and 45 °C 3 min. Hybridize overnight at 37 °C in a humid chamber (*see* **Note 7**). The hybridization solution for repetitive DNA probes has a stringency of 76 % at 37 °C (*see* **Note 8**).

*3.4.2 Post-hybridization Washing*

1. Remove the cover slips carefully in 2× SSC pre-warmed to 37 °C and wash the preparations at 79 % stringency (*see* **Note 8**) in 20 % formamide in 0.1× SSC twice for 5 min at 42 °C.

2. Remove the washing solution by rinsing for 5 min in 2× SSC twice at 42 °C and once at 37 °C.

*3.4.3 Antibody Reaction*

This step is necessary to remove indirect labeling with biotin or digoxigenin.

1. Equilibrate the slides in 4× SSC/0.2 % v/v Tween 20 for 5 min at 37 °C.

2. Apply 200 µL of 5 % BSA (bovine serum albumin) in 4× SSC/0.2 % Tween 20 per slide and incubate the preparations under plastic cover slips for 30 min at 37 °C in a humid chamber.

3. Apply 50 µL of the streptavidin and anti-digoxigenin dilution in 3 % BSA in 4× SSC/0.2 % Tween 20 per slide and incubate the slides under plastic cover slips for 1 h at 37 °C in a humid chamber (*see* **Notes 9** and **10**).

4. After the detection, wash off excess antibody three times in 4× SSC/0.2 % Tween at 37 °C.

5. Finally, apply 10 µL of 2 µg/mL DAPI solution and 10 µL of an anti-fading solution, and cover the preparations with glass cover slips and store at 4 °C.

**3.5 UV Microscopy**

Examine the slides with a fluorescent microscope equipped with the corresponding filter sets at 63× magnification (Fig. 4). Take care to photograph as fast as possible to avoid damage of fibers by UV light (*see* **Note 11**).

**Fig. 4** Photomicrograph of fiber FISH under an UV microscope. Two repetitive DNA probes were labeled with biotin and digoxigenin and then detected with streptavidin-Cy3 (*red*) and anti-digoxigenin-FITC (*green*), respectively. The length standard allows the measurement of array lengths

## 4   Notes

1. Depending on plant species, different types of material can be used. However, in most cases etiolated young tissue without pigments (seedlings or growing roots) will give the best yield and quality of nuclei.

2. The nuclei suspension should not be too dense (i.e., should not stick to each other) to allow optimal stretching of the chromatin. A portion of the nuclei may have already lysed, but there should not be too many damaged nuclei. Most nuclei should be round and evenly colored with DAPI.

3. The glass cover slip should slide down easily when evenly tilted at the degree of 30°–45°. If the glass does not slide, the STE volume and incubation time can be increased.

4. The amplification temperature varies depending on the primers' base composition. The amplification temperature is calculated according to the following formula:
    For sequences less than 14 nucleotides the formula is
    $$Tm = (wA + xT) \times 2 + (yG + zC) \times 4$$
    where w, x, y, and z are the numbers of the bases A, T, G, and C in the sequence, respectively.

For sequences longer than 13 nucleotides, the equation used is

$$Tm = 64.9 + 41 \times (yG + zC - 16.4)/(wA + xT + yG + zC)$$

5. The DNA labeled by PCR or nick translation is purified from unincorporated label by adding 2.5 volumes of ice-cold absolute ethanol and 0.1 volume of 1 M LiCl, followed by incubation at −20 °C for 3 h or −80 °C for 30 min and centrifugation for 20 min at 7,515 g in a microcentrifuge at 4 °C. The probes are then resuspended in an appropriate amount (typically 5–20 μL) of distilled sterile water.

6. These reagents are light sensitive. Add to buffer shortly before use and keep in the dark afterward.

7. Take utmost care that the slides do not dry out and remain wet during the whole procedure, unless you perform controlled dehydration in ethanol series (70 %, 90 %, 100 %).

8. The stringencies vary depending on the nature and composition of the probe and on the purpose of the experiment. Typically, start with the hybridization stringency of 76 % and washing stringency of 79 %. For stringency calculation, *see* ref. 16.

9. Antibody dilutions may vary depending on the copy number of the target sequence and the probe labeling efficiency. Typically, dilute using 3 % BSA in 4× SSC/0.2 % Tween. For digoxigenin labeled probes at 1:75 dilution anti-DIG-FITC and for biotin-labeled probes 1:200 dilution of streptavidin-Cy3 are recommended.

10. If the signals are too weak, use an antibody cascade to amplify the signals. Apply 50 μL of the appropriate antibody dilution in 3 % BSA in 4× SSC/0.2 % Tween 20 per slide and incubate the preparations under plastic cover slips. Wash after every step three times for 5 min in 4× SSC/0.2 % Tween at 37 °C.

*Detection of digoxigenin-labeled probes*

Step 1: Anti-DIG-alkaline phosphatase (e.g., Roche) 1:500 for 1 h at 37 °C.

Step 2: Fast Red detection solution (e.g., Roche) 1:5000 for 1 h at 37 °C.

*Detection of biotin- labeled probes*

Step 1: Streptavidin-horseradish peroxidase (Probes) 1:100 for 30 min at RT.

Step 2: Tyramide signal amplification detection solution (Probes) for 5 min at RT.

11. If the fibers are of good quality, they will be barely visible in DAPI and are quickly damaged by the UV light. Microscopy with DAPI filters should be avoided. It is helpful only to search the tracks of fluorescent signals using appropriate filters depending on the wavelengths of the applied fluorochromes.

## Acknowledgments

We gratefully acknowledge funding by the BMBF GABI-FUTURE START (grant 0315070), AnnoBeet project (0315962C). We are grateful to Prof. Dr. K. Fukui (Graduate School of Engineering, Osaka University, Osaka, Japan) and Prof. Dr. N. Ohmido (Graduate School of Human Development and Environment, Kobe University, Kobe, Japan) and their colleagues for the opportunity to perform fiber FISH in their laboratories and to M. Richter-Babekoff, International Office of the Technische Universität Dresden, for excellent administrative support.

## References

1. John HA, Birnstiel ML, Jones KW (1969) RNA-DNA hybrids at the cytological level. Nature 223:582–587

2. Manuelidis L, Langer-Safer PR, Ward DC (1982) High-resolution mapping of satellite DNA using biotin-labeled DNA probes. J Cell Biol 95:619–625

3. Rayburn AL, Gill BS (1985) Use of biotin-labeled probes to map specific DNA sequences on wheat chromosomes. J Hered 76:78–81

4. Schwarzacher T, Heslop-Harrison JS (1991) *In situ* hybridization to plant telomeres using synthetic oligomers. Genome 34:317–323

5. Heslop-Harrison JS, Schwarzacher T (2011) Organization of the plant genome in chromosomes. Plant J 66:18–33

6. Desel C, Jansen R, Dedong G, Schmidt T (2002) Painting of parental chromatin in Beta hybrids by multi-colour fluorescent in situ hybridization. Ann Bot 89:171–181

7. Fonsêca A et al (2010) Cytogenetic map of common bean (*Phaseolus vulgaris* L.). Chromosome Res 18:487–502

8. Ohmido N et al (2010) Integration of cytogenetic and genetic linkage maps of *Lotus japonicus*, a model plant for legumes. Chromosome Res 18:287–299

9. Paesold S, Borchardt D, Schmidt T, Dechyeva D (2012) A sugar beet (*Beta vulgaris* L.) reference FISH karyotype for chromosome and chromosome-arm identification, integration of genetic linkage groups and analysis of major repeat family distribution. Plant J 72:1–12

10. Fransz PF, Alonso-Blanco C, Liharska T, Peeters AJM, Zabel P, De Jong JH (1996) High resolution physical mapping in *Arabidopsis thaliana* and tomato by fluorescence in situ hybridisation to extended DNA fibres. Plant J 9:421–430

11. De Jong JH, Fransz P, Zabel P (1999) High resolution FISH in plants-techniques and applications. Trends Plant Sci 4:258–263

12. Ohmido N, Kijima K, Ashikawa I, de Jong JH, Fukui K (2001) Visualization of the terminal structure of rice chromosomes 6 and 12 with multicolor FISH to chromosomes and extended DNA fibers. Plant Mol Biol 47:413–421

13. Szinay D et al (2008) High-resolution chromosome mapping of BACs using multi-colour FISH and pooled-BAC FISH as a backbone for sequencing tomato chromosome 6. Plant J 56:627–637

14. Dechyeva D, Schmidt T (2006) Molecular organization of terminal repetitive DNA in *Beta* species. Chromosome Res 14:881–897

15. Paesold S (2013) Entwicklung eines FISH-Referenzkaryotyps für die Zuckerrübe (*B. vulgaris*) und Anwendungsmöglichkeiten in der Chromosomen- und Genomanalyse. PhD Thesis, Technische Universität Dresden, Dresden

16. Schwarzacher T, Heslop-Harrison JS (2000) Practical in situ hybridization. BIOS Scientific Publishers Limited, Oxford

17. Schmidt T, Schwarzacher T, Heslop-Harrison JS (1994) Physical mapping of rRNA genes by fluorescent *in situ* hybridization and structural analysis of 5S rRNA genes and intergenic spacer sequences in sugar beet (*Beta vulgaris*). Theor Appl Genet 88:629–636

# Chapter 4

# Tyramide Signal Amplification: Fluorescence In Situ Hybridization for Identifying Homoeologous Chromosomes

**Araceli Fominaya, Yolanda Loarce, Juan M. González, and Esther Ferrer**

## Abstract

Tyramide signal amplification (TSA) fluorescence in situ hybridization (FISH) has been shown as a valuable molecular tool for visualizing specific amplified DNA sequences in chromosome preparations. This chapter describes how to perform TSA-FISH, paying special interest to its two critical steps: probe generation and metaphase plate generation. The potential of physically mapping 12S-globulin sequences by TSA-FISH as a means of identifying homeology among chromosome regions of *Avena* species was tested and is discussed.

**Key words** TSA-FISH, Obtaining probes, Metaphase chromosomes, Physical mapping, *Avena*

## 1 Introduction

Fluorescence in situ hybridization (FISH) allows researchers to use fluorescent signals to identify chromosome-specific sequences, chromosomal segments, or whole sets of chromosomes. Conventional FISH involves the labeling of DNA with a reporter molecule, followed by the hybridization of the probe and target DNA, and then incubation with immunofluorescent reagents [1]. However, its use is limited by inadequate detection sensitivity for small genomic targets of low copy number. Tyramide signal amplification (TSA)-FISH, however, is reported to increase this sensitivity by up to 1000-fold [2, 3].

TSA (sometimes called CARD for "catalyzed reported deposition") is an enzyme-mediated detection method that uses the catalytic activity of horseradish peroxidase to induce high-density labeling of a target nucleic acid sequence [4]. TSA-FISH is a multistep procedure consisting of in situ hybridization with a biotinylated probe, the detection of the hybridized target with streptavidin-horseradish peroxidase, and signal amplification with fluorescent dye-conjugated tyramide (Fig. 1). Amplification is

Shahryar F. Kianian and Penny M.A. Kianian (eds.), *Plant Cytogenetics: Methods and Protocols*,
Methods in Molecular Biology, vol. 1429, DOI 10.1007/978-1-4939-3622-9_4,
© Springer Science+Business Media New York 2016

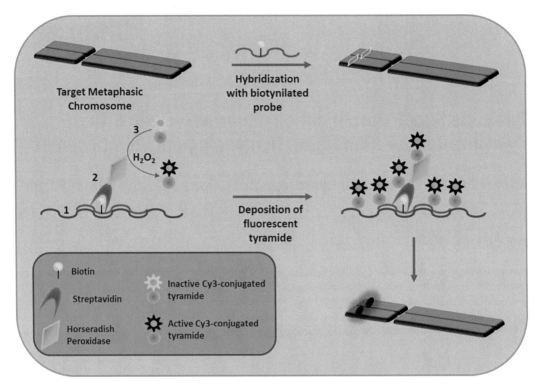

**Fig. 1** General overview of TSA-FISH detection of metaphase chromosomes

greatly increased by the use of this latter complex; it binds to the streptavidin-horseradish peroxidase, which in turn catalyzes the binding of the tyramide to the surface of the chromosomes on the slide. The key to TSA-FISH is to achieve ample deposition of this fluorescent dye-conjugated tyramide, amplifying the fluorescent signal [5]. The success of this procedure depends on two critical steps [6]: (1) optimizing the quality of the sequence used as the probe—a probe made with a heavily labeled mixture of short fragments spanning around 3000 bp facilitates the penetration of the condensed chromosomes allowing for highly efficient amplification of the hybridization signals; (2) optimizing the quality of the chromosome preparations—in metaphase plates, the absence of cytoplasm is crucial if strong contrast between the hybridization signal and the condensed chromosome is to be achieved.

This method can be used for the identification of homologs and homeologous chromosomes. TSA-FISH mapping of 12S-globulin (12S-Glob) in metaphase plates of diploid [*Avena strigosa* (AA genome), *A. eriantha* (CC genome)] and hexaploid [*A. sativa* cv. Ogle (AACCDD genome)] oat species (Fig. 2) was performed to establish relationships among the chromosomes bearing hybridization signals. The diploid species *A. strigosa* and *A. eriantha* are the putative donor species of the A and C genome of the hexaploid oat *A. sativa*, respectively [7, 8]. Using a 12S-Glob probe, TSA-FISH returned four hybridization signals at opposite

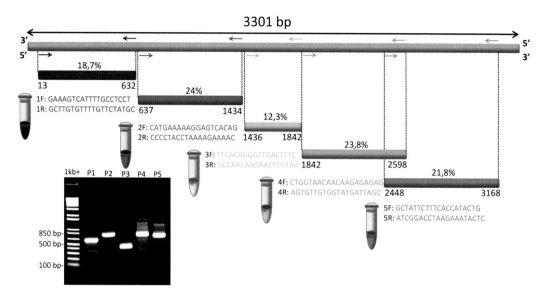

**Fig. 2** Strategy followed for primers designing from the sequence the 12S-globulin seed storage protein gene (GenBank accession J05485.1) for use in PCR amplification. *Thick-colored lines* span the total length of each PCR fragment. The numbers under the *colored lines* indicate the relative position of each fragment in the entire sequence. The numbers above the *line* indicate the proportion of the entire sequence occupied by the fragments. The photograph is an agarose gel with the PCR amplification result for each fragment

ends of a pair of chromosomes (labeled as no. 4) of *A. strigosa* (Fig. 2). In *A. eriantha*, one chromosome pair (labeled as no. 2) showed four hybridization signals on the same arm (Fig. 2). In *A. sativa* cv. Ogle, 12 hybridization signals were located interstitially on the long arms of the five chromosome pairs (Fig. 4c). These observations agree with the reported homologous and homeologous relationships among the chromosome regions of *Avena* species [6, 11]. These results show the usefulness of TSA-FISH when attempting to identify homologous and homeologous chromosomes in diploid and polyploid species.

## 2 Materials

### 2.1 Probe Generation

#### 2.1.1 Probe DNA Isolation, Electrophoresis, and Gel Extraction

1. Genomic DNA extraction kit for high-quality DNA (e.g., DNeasy Plant Mini Kit from Qiagen).

2. PCR Master Mix containing *Taq* DNA polymerase, dNTPs, plus all the other components required for PCR (e.g., Promega's Master Mix).

3. 1× TAE (Tris–HCl-acetate-EDTA) running buffer: 40 mM Tris, 20 mM acetic acid, and 1 mM EDTA (*see* **Note 1**).

4. Ethidium bromide: 0.625 mg/ml (*see* **Note 2**).

5. 10× gel loading buffer: 0.2 % bromophenol blue dye, 0.2 % xylene cyanol dye, and 30 % glycerol in Tris-EDTA buffer.

6. Molecular weight marker: 1 kb Plus DNA Ladder.

7. DNA gel extraction kit (e.g., QIAquick from Qiagen).

*2.1.2 Probe Labeling*

1. Deoxyribonuclease I 20,000 U/µl (*see* **Note 3**).

2. DNA polymerase I 5 U/µl.

3. 1 mM Biotin-16-dUTP.

4. 100 mM Deoxynucleotide set.

5. 10× Nick translation buffer: 500 mM Tris–HCl, pH 7.8, 50 mM $MgCl_2$, 100 mM 2-mercaptoethanol, bovine serum albumin 100 µg/ml.

6. 2× Deoxyribonuclease I dilution buffer: 40 mM NaAc pH 6.5, 10 mM $CaCl_2$, 0.2 mM phenylmethylsulfonyl fluoride (PMSF), 50 % glycerol.

*2.1.3 Solutions for DNA Concentration, Washing, and Preservation*

1. 1 M Tris–HCl pH 8.

2. 0.5 M EDTA pH 8.

3. Tris-EDTA (TE) buffer solution: 10 mM Tris–HCl, 1 mM disodium EDTA, pH 8.0.

4. 70 % ethanol (v/v).

5. 100 % ethanol.

6. 3 M sodium acetate pH 5.2.

## 2.2 Generation of Chromosome Preparations for FISH Using TSA

*2.2.1 Metaphase Spreads*

1. 100 % Absolute ethanol.

2. Glacial acetic acid.

3. 10× Citric acid-sodium citrate buffer (pH 4.8): 40 ml 0.1 M citric acid monohydrate, 60 ml 0.1 M tri-sodium citrate-2-hydrate.

   Prepare by dissolving 0.36 g of citric acid monohydrate in 160 ml of $dH_2O$, and then in a second container, dissolve 1.17 g of tri-sodium citrate-2-hydrate in 40 ml of $dH_2O$; finally mix the two solutions together to obtain a 4.8 pH solution.

4. Enzyme solution is made by mixing 200 ml of 1× citric sodium buffer, 4 g of cellulase (e.g., Calbiochem),1 g of cellulase (e.g., Onozuka R-10), and 50 ml of pectinase. Divide it into 1.5 ml microcentrifuge tubes. Store at −20 °C.

*2.2.2 TSA-FISH*

1. Tyramide signal amplification (**TSA**) Plus Cyanine 3 Kit (e.g., Perkin Elmer, NEL-744001KT)—contains Cy3-tyramide and 1× Plus amplification diluent (*see* **Note 4**).

2. Dimethyl sulfoxide (DMSO) (*see* **Note 4**).

3. Streptavidin-horseradish peroxidase.

4. Blocking reagent: 5 % w/v bovine serum albumin.

5. 20× SSC: 3 M NaCl, 300 mM tri-sodium citrate, adjusted to pH 7 with HCl. This stock solution is used to make other SSC concentrations.

6. 4× SSC/0.2% Tween mix 200 ml of 20× SSC, 2 ml Tween, and 800 ml of dH$_2$O. Store at RT.

7. TNT buffer: 0.1 M Tris–HCl pH 7.5, 0.15 M NaCl, 0.05% Tween 20.

8. TNB blocking buffer: 0.1 M Tris–HCl pH 7.5, 0.15 M NaCl, 0.5% blocking agent (*see* **Note 5**).

9. RNase stock (10 mg/ml): Dissolve 10 mg of RNase in 1 ml of 10 mM Tris–HCl pH 7.5 and 15 mM NaCl. Boil for 15 min and allow to cool. Divide into 1.5 ml microcentrifuge tubes. Store at –20 °C.

10. Paraformaldehyde 4% (w/v): Dissolve 4 g of paraformaldehyde in 100 ml of dH$_2$O. Heat the solution to 60 °C until it becomes milky. Add 2 ml of 0.1 M NaOH.

11. Bovine serum albumin (BSA) 5% (w/v) in 4× SSC solution: Dissolve 1 g of BSA in 20 ml of 4× SSC/Tween. Divide among 1.5 ml microcentrifuge tubes. Store at –20 °C.

12. Ethanol at various concentrations: 70% ethanol, 90% ethanol, and 100% ethanol.

13. Plastic coverslips (18×18 mm).

14. 20% Formamide.

# 3    Methods

## 3.1    Probe Generation

### 3.1.1    PCR Primer Design

Labeling of high molecular weight DNA molecules using the nick translation protocol requires large amounts of starting DNA and is difficult to control the degree and consistency of final labeled fragments, especially when dealing with large genomic DNA fragments. In contrast, PCR ensures more homogeneous labeling of the sequence and allows more accurate measurement of the probe quantities used. However, labeling by PCR works less efficiently with high molecular weight sequences. Thus, for labeling sequences over 1500 bp in length, several internal pairs of primers should be designed and used in independent PCR runs. A set of labeled fragments of the target sequences that might partially overlap is thus obtained.

Primers should be designed from the desired genome sequence using one of the online primer design software options available. Primers amplifying fragments (400–1000 bp) spanning the entire genome sequence are required. The critical conditions for primer design are length of 18–24 bases, 40–60% G-C content, minimum intra-primer and inter-primer homology (no more than 3 bp), and few to no stretches of polynucleotide repeats.

*3.1.2 Genomic DNA Extraction*

Genomic DNA is extracted from young test leaves using a plant DNA extraction kit (e.g., DNeasy Plant Mini Kit from Qiagen).

1. Freeze 100 mg of leaves in a mortar with liquid nitrogen and grind with a pestle.

2. Extract the DNA from the homogenate following the kit's instructions.

3. Elute the DNA with 100 µl of elution buffer provided by the kit or in TE.

4. Quantify the extracted DNA (*see* **Note 6**).

*3.1.3 PCR Amplification*

The pairs of designed primers are used in PCR amplifications of 50 ng genomic DNA, employing the PCR Master Mix.

1. Introduce into the PCR tubes all the components of the reactions: 12.5 µl of PCR Master Mix, 1 µl of gDNA, 10 µM of each forward primer and reverse primer, and add $ddH_2O$ up to 25 µl.

2. Place the tubes in the thermocycler. The general reaction conditions for any pair of primers are 40 cycles of 30 s at 95 °C, 30 s at 5 °C below the primer pair's Tm for primer annealing, and 30–60 s at 72 °C.

*3.1.4 Agarose Electrophoresis*

1. Prepare a 1 % (w/v) agarose gel in 1× TAE buffer.

2. Mix the PCR products with loading buffer and load the samples into the gel. Include a molecular weight marker in a separate well to guide in size selection of PCR product.

3. After the adequate separation of DNA fragments, stop gel electrophoresis.

4. Stain the gel in an ethidium bromide solution for 15 min (*see* **Note 2**).

5. Place the gel on a UV transilluminator to visualize the amplified DNA fragments (*see* **Note 7**).

6. While working on a UV transmitter, cut slices of the gel with the desired DNA fragment using a sharp scalpel and put them in a microcentrifuge tube.

*3.1.5 DNA Extraction from the Agarose Gel*

1. Extract the DNA from the agarose using a simple kit (e.g., QIAquick from Qiagen) following the manufacturer's instructions.

2. Elute the DNA with 100 µl of elution buffer or TE.

3. Precipitate the DNA with 1/10 vol 3 M sodium acetate pH 5.2 and two volume of absolute ethanol. Leave the 1.5 ml microcentrifuge tubes in the freezer at −20 °C for at least 20 min.

4. Centrifuge for 15 min at $20,000 \times g$ at 4 °C in a microcentrifuge; discard the supernatant.

5. Wash salts from the pellet with 70 % ethanol. Centrifuge for 10 min at 20,000 × $g$ at room temperature (RT); discard the supernatant.

6. Resuspend the DNA in 10 μl of dH$_2$O.

7. Quantify the concentration of the gel extracted PCR band.

*3.1.6 Nick Translation Probe Labeling*

1. 2 μg total of the DNA amplified from the various primer set is used in each labeling reaction.

2. Calculate the relative proportion of each PCR fragment spanning the probe (*see* **Note 8**). This allows for estimation of the relative amount of DNA from each PCR fragment in the 2 μg of probe needed for labeling (Fig. 2).

3. Add 2 μg total of combined DNA to a PCR tube with 5 μl of 10× nick translation buffer, 1 μl of 1 mM biotin-16-dUTP, and 2 mM dNTPs, and add dH$_2$O to 34.6 μl. Mix by pipetting.

4. Add 16 μl of DNA pol I (5 U/μl) and 0.4 μl of DNAse I (100 mU/μl) (*see* **Note 3**). Incubate for 2 h at 15 °C. Mix by pipetting.

5. Stop the reaction with 1 μl of 0.5 M EDTA pH 8 for 10 min at 65 °C.

6. Precipitate the probe with two volume of absolute ethanol, and then wash away the salts with 70 % ethanol. Proceed as in Subheading 3.1.5, **steps 4** and **5**.

7. Resuspend the DNA in 40 μl of ddH$_2$O.

8. Load 2 μl onto a 1.5 % agarose gel to check the size of the fragments generated after labeling (about 200 bp).

*3.2 Preparation of FISH Chromosome Spreads*

Actively growing roots can be collected from germinating seeds in a Petri dish.

*3.2.1 Collection of Root Tips*

1. Place the seeds on a moistened filter paper.

2. Keep the Petri dish with seeds in an oven at 25 °C for 48 h.

3. Transfer the Petri dish with the seeds to a dark cold room or refrigerator (0–4 °C) for 3–5 days.

4. Remove the Petri dish with the seeds from the refrigerator and leave at 25 °C for 24 h.

5. Cut only the meristematic region of the roots and drop into a microcentrifuge tube containing cold water; surround the tube in abundant ice. Keep the container in a refrigerator for 24 h.

6. Transfer the roots to a new microcentrifuge tube containing absolute ethanol-glacial acetic acid (3:1) freshly made and keep at room temperature for 2–3 days. Replace the ethanol-glacial acetic acid (3:1) and keep at −20 °C until use.

*3.2.2  Chromosome Preparation*

1. Place the fixed roots (Subheading 3.2.1, **step 6**) in citric acid-sodium citrate buffer, pH 4.8, for 20–25 min with agitation. Then change the buffer for a fresh buffer and leave for 1–3 h with agitation.

2. Transfer the roots into enzyme solution. Incubate at 37 °C for 1.5–2 h.

3. Wash the roots in citric acid-sodium citrate buffer for at least 20 min.

4. Transfer the material to a slide with a drop of 60% (v/v) acetic acid in water and excise the root meristem with forceps. Remove all tissue debris, cover with a 18 × 18 mm coverslip, squash cells, and observe. Slides deemed useful for further use should have a minimum of five cells at the metaphase stage with well-spread chromosomes.

5. Freeze the slide and remove the coverslip with a scalpel blade by prying it away from the slide.

6. Leave the slides to air dry at room temperature for 24–72 h.

### 3.3  TSA-FISH

*3.3.1  Pretreatment of Chromosome Preparations*

1. Use a 1:100 dilution of RNase in 2× SSC stock solution. Add 100 μl to each slide and cover with a plastic coverslip.

2. Incubate the slides in a humidity chamber (Fig. 3) at 37 °C for 1 h.

3. In a Coplin jar, wash the slides with 2× SSC twice for 5 min at RT with agitation.

4. Transfer the slides to another Coplin jar containing 4% paraformaldehyde solution at RT, and agitate for 10 min.

**Fig. 3** Humidity chamber (set on an agitator) for TSA-FISH signal—(**a**) open and (**b**) closed. The numbers indicate: *1*. microscope slide placed horizontal; *2*. plastic coverslip; *3*. slide holder to elevate slide over paper towel; *4*. wet filter paper; *5*. chamber; *6*. shaker; *7*. closed moist chamber

5. Wash the slides with 2× SSC twice for 5 min each at RT with agitation.

6. Incubate the slides in 70, 90, and 100 % ethanol for 3 min each with agitation.

7. Leave the slides to air dry at RT.

*3.3.2 Hybridization Mixture*

Prepare the hybridization solution by mixing all the required reagents (stored at –20 °C). Use a final volume of 30 μl/slide.

1. Introduce the components of the mixture—15 μl of 100 % formamide, 6 μl of 50 % dextran sulfate, 3 μl of 20× SSC, 1 μl of 10 % SDS, 1 μl of salmon sperm DNA and the DNA probe—into a 1.5 ml microcentrifuge tube. The probe concentration should be 150–200 ng per slide. Adjust the volume to 30 μl with ddH₂O. Mix strongly with a vortex and then spin in a microcentrifuge.

2. Denature the probe mixture at 80 °C for 15 min and then incubate on ice for 5 min.

*3.3.3 Denaturation and Hybridization*

1. Place the slides in a programmable thermal controller. Use this program: 75 °C for 7 min, 55 °C for 2 min, 50 °C for 30 s, 45 °C for 1 min, 42 °C for 2 min, 40 °C for 5 min, and 38 °C for 5 min, and finish at 37 °C.

2. When the program is finished, incubate the slides in a humidity chamber at 37 °C overnight to allow hybridization.

*3.3.4 Washing and Detection of the Amplified Hybridization Signal*

1. Wash the slides with 2× SSC for 5 min at 42 °C with agitation.

2. Wash the slides with 20 % formamide for 10 min at 42 °C with agitation.

3. Wash the slides with 0.1× SSC for 5 min at 42 °C with agitation.

4. Wash the slides with 2× SSC for 5 min at 42 °C with agitation.

5. Wash the slides with 4× SSC/Tween for 5 min at 42 °C with agitation.

6. Wash the slides with TNT buffer for 5 min at RT with agitation.

7. Add 100 μl TNB per slide, cover with plastic coverslip, and incubate in a humidity chamber for 30 min at RT.

8. Add 100 μl conjugated streptavidin-HRP diluted in TNB blocking buffer (1:1000 dilution) per slide. Cover slides with plastic coverslips and incubate in a humidity chamber for 40 min at RT.

9. Wash the slides for 5 min in TNT buffer at RT with agitation, repeat twice. Keep slides in a Coplin jar.

10. Shake off any excess buffer, apply 100 µl tyramide solution diluted in amplification diluent (1:50 dilution) per slide, and cover with a plastic coverslip. Incubate the slides in a humidity chamber for 7 min at RT with gentle agitation.

11. Wash the slides for 5 min in TNT buffer at RT with agitation, repeat twice. Keep slides in a light-tight Coplin jar.

12. Shake off excess buffer, apply 100 µl DAPI solution per slide, and cover slide with a plastic coverslip. Incubate the slides for 15 min in a cover box at RT.

13. Wash the slides quickly in 4× SSC/Tween at RT.

14. Apply 35 µl of antifade to each and cover with a 24 mm × 60 mm coverslip.

15. Slides are ready to be viewed with fluorescent microscope.

# 4    Notes

1. Usually a 50× TAE stock solution is made from which 1× TAE buffer is prepared with dH$_2$O.

2. Ethidium bromide is a toxic chemical; handle with care and wear protective clothing and gloves. For gel staining, add five drops of ethidium bromide solution to 250 ml of dH$_2$O.

3. The DNAse enzyme (20,000 U/µl) must be diluted to the working concentration (100 mU/µl) with 2× deoxyribonuclease I dilution buffer.

4. Add 150 µl DMSO to the Cy3-tyramide and store at 4 °C.

5. Add the blocking reagent slowly in small increments while stirring. Heat gradually to 60 °C with continuous stirring to completely dissolve the blocking reagent. Prepare aliquots and store at –20 °C for long-term use.

6. A nanodrop spectrophotometer will measure the concentration and purity of the extracted DNA.

7. UV light causes damage to the eyes and skin; use a mask. Avoid long exposure of the gel to UV light since this will degrade the DNA fragments.

8. TSA-FISH mapping of 12S-globulin (12S-Glob) in metaphase plates of diploid *A. strigosa* (AA genome), *A. eriantha* (CC genome), and hexaploid *A. sativa* cv. Ogle (AACCDD genome) oat species (Fig. 4) was performed to establish relationships among the chromosomes bearing

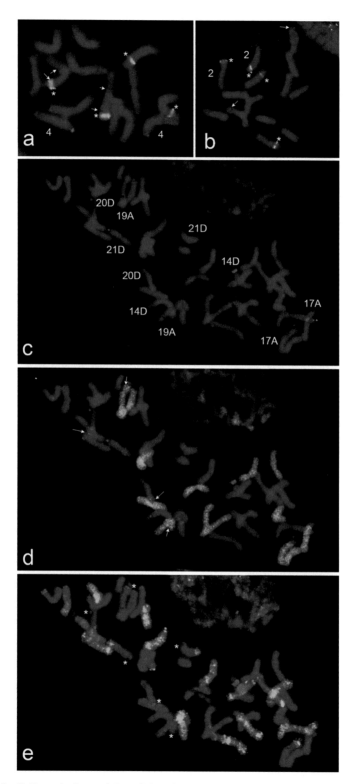

**Fig. 4** TSA-FISH of mitotic metaphase plates of *A. strigosa* (**a**), *A. eriantha* (**b**), and *A. sativa* cv. Ogle (**c**) using the biotin-labeled 12S-Glob probe (*red*). Chromosomes are counterstained with DAPI. The identified chromosomes are indicated by *numbers. Asterisks* indicate NOR-bearing chromosome pairs after rehybridization with digoxigenin-labeled pITS probe (**a**, **b**) or biotin-labeled pTa71 probe (**e**). *Arrowheads* indicate 5S-bearing chromosome pairs after rehybridization with the biotin-labeled pTa794 probe (**a**, **b**, and **d**). *See* **Note 8** for further discussion of results from these figures

hybridization signals. The diploid species *A. strigosa* and *A. eriantha* are the putative donor species of the A and C genome to the hexaploid oat *A. sativa* [7, 8]. Using a 12S-Globulin probe, TSA-FISH returned four hybridization signals on the chromosomes of *A. strigosa* (Fig. 4a). This chromosome pair was identified by size and arm ratio as As4 [9]. In *A. eriantha*, one chromosome pair showed four hybridization signals, located in telomeric and interstitial regions of the long arm (Fig. 4b). This chromosome pair was identified as Cp2 by the presence of two previously described ITS loci [9]. In *A. sativa* cv. Ogle, 12 hybridization signals were located interstitially on the long arms of five chromosome pairs (Fig. 4c).

Subsequent conventional FISH with the repeated genome-specific probes (120a [A genome specific] and Am1 [C genome specific]) and the ribosomal probes (ITS and 794) identified the five chromosome pairs as 17A, 19A, 14D, 20D, and 21D. Based on the hypothesis that the D genome of the hexaploid species originated from a species closely related to the A genome diploid species [8], the present observations suggest that all these hexaploid chromosome pairs are related to chromosome pair As4.

Chromosome 17A is a translocated chromosome. The terminal portion of its long arm carries C genome chromatin belonging to chromosome 7C [10, 11]. 17A-7C showed two 12S-Glob hybridization signals (Fig. 4c–e). One was located on the C genome chromatin and therefore derived from chromosome 7C. Chromosome 21D is also a translocated chromosome. The interstitial region of its long arm contains C genome chromatin belonging to an unknown chromosome [11]. Analysis of the relative signal intensities in individual chromosome pairs showed the hybridization signals on chromosome pairs 17A–7C and 14D to be generally more intense than those on 19A, 20D, and 21D. Together, these observations suggest that chromosome pairs 17A–7C and 14D are related to the long arm of chromosome pair As4, whereas chromosome pairs 19A, 20D, and 21D are related to the short arm of chromosome pair As4 (Fig. 5). The TSA-FISH mapping of resistance gene analogs to hexaploid species revealed chromosome pairs 17A–7C and 14D to show hybridization signals with the III2.18-RGA probe, whereas chromosome pairs 19A and 20D showed hybridization signals with II2.17-RGA.

These results show the usefulness of TSA-FISH when attempting to identify homologous and homeologous chromosomes in diploid and polyploid species.

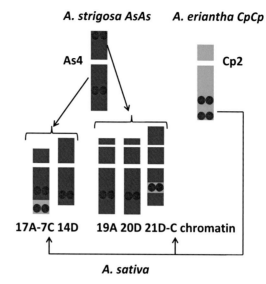

**Fig. 5** Diagram showing relationships among chromosomes from diploid and hexaploid species based on hybridization patterns of chromosomes with 12S-Glob sequences

## Acknowledgments

We thank the *Ministerio de Ciencia e Innovación* of Spain (AGL210-17042) and *Universidad de Alcalá* (UAH GC2014-002 and UAH CCG2014/EXP-068) for its support of this work.

## References

1. Leich AR, Schwarzacher T, Jackson D, Leitch IJ (1994) In situ hybridization. BIOS Scientific Publishers Limited, Oxford

2. Bobrow MN, Litt GJ, Shaughnessy KJ, Mayer PC, Conlon J (1992) The use of catalysed reported deposition as a means of signal amplification in a variety of formats. J Immunol Methods 150:145–149

3. Raap AK (1998) Advances in fluorescence in situ hybridization. Mutat Res 400:287–298

4. Bobrow MN, Harris TD, Shaughnessy KJ, Litt GJ (1989) Catalyzed reported deposition, a novel method of signal amplification. Application to immunoassays. J Immunol Methods 125:279–285

5. Schriml LM, Padilla-Nash HM, Coleman A, Moen P, Nash WG, Menninger J, Jones G, Ried T, Dean M (1999) Tyramide signal amplification (TSA)-FISH applied to mapping PCR-labeled probes less than 1 kb in size. Biotechniques 27:608–613

6. Sanz MJ, Loarce N, Ferrer E, Fominaya A (2012) Use of tyramide-fluorescence in situ hybridization and chromosome microdissection for ascertaining homology relationships and chromosome linkage group associations in oats. Cytogenet Genome Res 136:145–156

7. Thomas H (1992) Cytogenetics of *Avena*. In: Marshall HG, Sorrells ME (eds) Oat science and technology, vol 33, Agron Monogr. ASA, CSSA, Madison, WI, pp 473–507

8. Linares C, Ferrer E, Fominaya A (1998) Discrimination of the closely related A and D genomes of the hexaploid oat Avena sativa L. Proc Natl Acad Sci U S A 95:12450–12455

9. Linares C, González J, Ferrer E, Fominaya A (1996) The use of double fluorescence in situ hybridization to physically map the positions of 5S rDNA genes in relation to the chromosomal location of 18S–5.8S-26S rDNA and a C genome specific DNA sequence in the genus Avena. Genome 39:535–542

10. Jellen EN, Beard J (2000) Geographical distribution of a chromosome 7C and 17 intergenomic translocation in cultivated oat. Crop Sci 400:256–263

11. Sanz MJ, Jellen EN, Loarce L, Irigoyen ML, Ferrer E, Fominaya A (2010) A new chromosome nomenclature system for oat (*Avena sativa* L. and *A. byzantina* C. Koch) based on FISH analysis of monosomic lines. Theor Appl Genet 121:1541–1552

# Chapter 5

# Localization of Low-Copy DNA Sequences on Mitotic Chromosomes by FISH

## Miroslava Karafiátová, Jan Bartoš, and Jaroslav Doležel

## Abstract

Fluorescence in situ hybridization (FISH) is a widely used method to localize DNA sequences on mitotic and meiotic chromosomes and interphase nuclei. It was developed in early 1980s and since then it has contributed to numerous studies and important discoveries. Over the decades, the protocol was modified for ease of use, allowing for localizing multiple probes simultaneously and increasing its sensitivity and specificity. Despite the continuous improvements, the ability to detect short single-copy sequences of only a few kilobases or less, such as genes, remains limited. Here, we provide a detailed protocol for detection of short, single- or low-copy sequences on plant mitotic metaphase chromosomes.

**Key words** Cell cycle synchronization, Cytogenetic mapping, Fluorochrome, Low-copy probe, Mitotic metaphase chromosomes, Probe purification, Post-fixation, Quantum yield

## 1 Introduction

Knowing the physical position of DNA sequences within a genome is critical to understanding their structure, evolution, and function and essential in attempts to transfer the underlying genes to other genotypes and in cloning. Since the formulation of the chromosome theory of heredity in the beginning of the twentieth century, various approaches were used to infer the position of genes and later other DNA sequences along chromosomes and within interphase nuclei. They are broadly classified into two categories: genetic and physical. While the genetic linkage mapping estimates position of genetic elements based on the frequency of meiotic recombination, physical mapping determines real, physical position of DNA elements [1]. The ultimate map of a genome is the reference genome sequence in which the position of a DNA sequence is determined with precision to a single base pair. However, such sequences are available only for a few plant species [2–4], which were sequenced through the clone-by-clone approach.

Shahryar F. Kianian and Penny M.A. Kianian (eds.), *Plant Cytogenetics: Methods and Protocols*,
Methods in Molecular Biology, vol. 1429, DOI 10.1007/978-1-4939-3622-9_5,
© Springer Science+Business Media New York 2016

The advent of next-generation sequencing (NGS) technology revolutionized plant genetics and genomics, making whole genome shotgun sequencing a method to generate genome sequences at a lower cost [5–7]. However, while NGS excels in high throughput, the assembly of a genome sequence from short sequence reads remains a challenge especially in species with large and polyploid genomes characterized by enormous sequence redundancy. Sequence contigs are typically ordered using genetic markers, whose position in proximal chromosome regions is difficult to determine due to the poor resolution of genetic maps. Yet, such regions may span well over one third of a genome [8]. Clearly, additional approaches are needed to support and validate sequenced contig order. The use of fluorescence in situ hybridization (FISH) with probes for single-/low-copy sequences is one of the attractive approaches to accomplishing this goal [9–11].

In situ hybridization allows visualizing of DNA sequences directly onto the chromosomes and nuclei. The principle is based on the ability of the DNA sequence of interest (probe) to hybridize with high specificity to the complementary sequence on a target DNA under the strictly controlled conditions. In order to visualize the site(s) of probe hybridization, the probe must be labeled. Originally, DNA probes were conjugated with a radioactive isotope; however, this inconvenient approach was soon replaced by enzymatically and later fluorescently labeled probes. FISH is most frequently performed on mitotic metaphase chromosomes due to their ease of availability [12]. Nevertheless, other stages of mitosis and meiosis, which provide different degree of resolution, such as mitotic prometaphase [13], meiotic pachytene [14], or interphase nuclei [15], have been used. In order to achieve higher spatial resolution, FISH is done on longitudinally stretched flow-sorted chromosomes [16] and on DNA fibers released from interphase nuclei [17].

In contrast to genetic mapping, the resolution of cytogenetic maps is not compromised due to a low frequency of recombination in proximal regions [18]. In principle, any piece of chromosome DNA can be used for probe preparation and its position revealed by FISH [19–21]. However, the presence of dispersed repetitive DNA elements imposes limitations on probe selection. In genomes with a low portion of dispersed repeats, DNA clones from large-insert libraries (e.g., cloned in bacterial artificial chromosome, BAC), which are easy to localize due to their length, can serve as single-copy probes [22–25]. On the other hand, FISH with BAC clones from large genomes with prevalence of dispersed repeats results in dispersed hybridization signals preventing their unambiguous localization [26, 27]. The addition of unlabeled $C_0t$-1 fraction of genomic DNA to hybridization mix may reduce nonspecific hybridization signals [24], but this approach is usually not efficient in large genomes [28].

As large BAC clones are not useful as FISH probes in species with complex genomes, one solution is to use their parts (subclones) to lower the chance of hybridization with repeat sequences. However, this approach has not met with significant success as the subclones may still contain repetitive sequences [28]. If sequences of BAC clones are available, a solution is to identify single-copy genomic sequences and design primers to amplify short sequences suitable for FISH [10]. However, using sequences shorter than 10 kb to prepare FISH probes may reach the lower detection limits of FISH [17]. Generally, probes over 10 kb long are mapped routinely in plants [29, 30]. However, in complex genomes, unique motifs are restricted to only a few kb-long regions, usually corresponding to genes and genic sequences [8, 31]. Despite the progress in FISH methodology and fluorescence microscopy, localization of such small DNA regions still cannot be considered a routine. Reports on successful mapping of DNA sequences shorter than 5 kb on plant chromosomes by FISH are scarce [8–10, 31–34].

This protocol describes the localization of full-length cDNA (fl-cDNA) clones of limited length in barley by FISH. In order to increase the number of mitotic metaphase plates available for FISH, the protocol includes a procedure to induce cell cycle synchrony and accumulate synchronized cells in metaphase. This protocol can be modified to work with other plant species using the cell cycle synchronization described elsewhere in this book.

## 2 Materials

### 2.1 Plant Material

Vernalized seeds of barley (*Hordeum vulgare* L., cv. Morex)

### 2.2 Reagents and Solutions

#### 2.2.1 Reagents and Solutions for Cell Cycle Synchronization and Accumulation of Metaphases

1. Solution A: 45 mM $H_3BO_3$ (280 mg), 20 mM $MnSO_4 \cdot H_2O$ (340 mg), 0.4 mM $CuSO_4 \cdot 5H_2O$ (10 mg), 0.8 mM $ZnSO_4 \cdot 7H_2O$ (22 mg), and 0.08 mM $(NH_4)_6Mo_7O_{24} \cdot 4H_2O$ (10 mg) in deionized water (100 mL). Store at 4 °C.

2. Solution B: 0.05 mM concentrated $H_2SO_4$ (0.5 mL) in deionized water (100 mL). Store at 4 °C.

3. Solution C: 18 mM $Na_2EDTA$ (3.36 g) and 2.79 g $FeSO_4$ (20 mM) in deionized water. Heat the solution to 70 °C while stirring until the color turns yellow-brown. Cool down, adjust the volume with deionized water to 500 mL, and store at 4 °C.

4. Hoagland's stock solution (10×): 40 mM $Ca(NO_3)_2 \cdot 4H_2O$ (4.7 g), 20 mM $MgSO_4 \cdot 7H_2O$ (2.6 g), 65 mM $KNO_3$ (3.3 g), 10 mM $NH_4H_2PO_4$ (0.6 g), 5 mL solution A, and 0.5 mL solution B in deionized water. Adjust volume to 500 mL. Store at 4 °C for no more than 2 weeks.

5. Hoagland's nutrient solution (0.1×): 10 mL Hoagland's stock solution (10×) and 0.5 mL solution C in deionized water. Adjust volume to 1000 mL. Prepare just before use.

6. 2 mM hydroxyurea (HU) solution: dissolve 121.6 mg hydroxyurea in 800 mL 0.1× Hoagland's nutrient solution. Prepare just before use.

7. Amiprophos-methyl (APM) stock solution (20 mM): dissolve 60.86 mg APM in 10 mL ice-cold acetone and store at –20 °C, in 1 mL aliquots.

8. APM working solution (2.5 μM): 101.3 μL APM stock solution in 800 mL deionized water. Prepare just before use.

*2.2.2  Reagents and Solutions for Root Fixation and Slide Preparation*

1. 3:1 fixative: mix absolute ethanol and glacial acetic acid in 3:1 ratio. Prepare just before use.

2. 45 % acetic acid: mix 45 mL 99 % acetic acid and 54 mL ddH$_2$O.

3. 2 % carmine acid solution: 2 g carmine powder dilute in 100 mL 45 % acetic acid. After 30 min of mixing and heating, filter the suspension. Store in a dark glass bottle at RT.

4. 96 % ethanol.

*2.2.3  Reagents and Solutions for Preparation of FISH Probes*

1. 10× nick translation buffer: 50 mL 1 M Tris–HCl (pH 7.5), 5 mL 1 M MgCl$_2$, 50 mg BSA, and 45 mL ddH$_2$O. Make 100 μL aliquots and freeze at –20 °C.

2. 2 mM non-labeled dNTPs: dilute 100 mM dATP, dCTP, dGTP, and dTTP to 2 mM solutions (2 μL of stock and 98 μL ddH$_2$O) and then mix 100 μL 2 mM dATP, 100 μL 2 mM dCTP, 100 μL 2 mM dGTP, 20 μL 2 mM dTTP, and 80 μL dd H$_2$O. Make 100 μL non-labeled dNTP aliquots and store at –20 °C.

3. 0.1 M mercaptoethanol: 0.1 mL mercaptoethanol and 14.4 mL dd H$_2$O. Make aliquots and freeze at –20 °C.

4. Labeled dUTP: Texas Red-dUTP, biotin-dUTP and digoxigenin-dUTP, or DEAC-dUTP.

5. DNase I.

6. DNA polymerase I.

7. Input DNA: isolated and purified cDNA inserts originating from pericentromeric region of barley chromosome 7H. PCR reagents for cDNA insert amplification: 1 mM PCR buffer containing 1.5 mM MgCl$_2$, 0.2 mM of each dNTPs, 1 mM T3/T7 primers, and 2U/5 μL DNA polymerase in 50 μL.

8. TE (Tris EDTA) buffer: 10 mM Tris–HCl, pH 7.5. 1 mM EDTA pH 8.0.

9. Herring sperm DNA.

10. 3 M sodium acetate (NaAc): dissolve 40.8 g sodium acetate trihydrate in 70 mL of deionized water. Adjust the pH to 5.2

by adding HCl. Add water to bring the total volume of solution to 100 mL.

11. Chilled 96 and 70 % ethanol.

*2.2.4   Reagents and Solution for Post-fixation and FISH*

1. 20× SSC stock solution: 3 M NaCl (175.3 g) and 300 mM $Na_3C_6H_5O_7 \cdot 2H_2O$ (88.2 g) in deionized $H_2O$ (1000 mL). Adjust pH to 7. Sterilize by autoclaving. Store at room temperature.

2. 2× SSC washing buffer: 20× SSC (100 mL) in deionized $H_2O$ (900 mL).

3. 0.1× SSC stringent washing buffer: 20× SSC (5 mL), 0.1 % Tween 20 (1 mL), and 2 mM $MgCl_2 \cdot 6H_2O$ (406 mg) in deionized $H_2O$ (1000 mL).

4. 4× SSC washing buffer: 20× SSC (200 mL) and 0.2 % Tween 20 (2 mL) in deionized $H_2O$ (1000 mL).

5. 45 % acetic acid (*see* Subheading 2.2.2).

6. 4 % formaldehyde solution: add 5 mL 37 % formaldehyde into 42 mL 2× SSC.

7. 70, 96, and 100 % ethanol dehydration series.

8. 4× buffer (200 µL): 20× SSC (80 µL), 1 M Tris–HCl pH 8.0 (8 µL), 0.5 M EDTA (1.6 µL), herring sperm (10 µg/µL ssDNA, 11.2 µL), and $ddH_2O$ (99.2 µL). Make 50 µL aliquots and store at –20 °C.

9. Deionized formamide.

10. Hybridization mix: 50 % formamide (10 µL), 4× buffer (5 µL), and labeled DNA probe(s) (300 ng/µL). Add $ddH_2O$ up to 20 µL final volume. Prepare just before use. Labeled DNA probes (either directly labeled with fluorescent probes or labeled by digoxigenin or biotin) may be prepared using nick translation according to [8].

11. Two-layer detection of digoxigenin-labeled probes: FITC-labeled anti-digoxigenin antibody raised in sheep and anti-sheep FITC antibody.

12. Three-layer detection of biotin-labeled probes: Cy3-labeled streptavidin antibody, biotinylated anti-streptavidin, and Cy3-labeled streptavidin.

13. 1× blocking solution: dissolve 0.5 g blocking reagent in 50 mL 4× SSC. Autoclave. Store at –20 °C, in 1 mL aliquots.

14. Vectashield antifade mounting medium containing DAPI.

*2.3   Laboratory Devices and Other Equipment*

1. Plastic boxes (750 mL) including plastic cover with drilled holes (1–3 mm in diameter).

2. Aquarium aerating system with air stones.

3. Fluorometer or spectrophotometer for DNA concentration measurement.

4. Microscopic slides and cover slips.

5. Thermal cycler.

6. Humidity chamber at 37 °C.

7. Compound light microscope.

8. Fluorescence microscope equipped with filter blocks for Cy3, FITC, DAPI, DEAC, Texas Red, digital camera, and appropriate imaging system.

9. Image capturing software.

10. Rubber cement.

## 3   Methods

### 3.1   Seed Germination, Cell Cycle Synchronization, and Metaphase Accumulation (See Note 1)

1. Soak the seeds in ddH$_2$O for 15 min at room temperature (RT).

2. Transfer the seeds into glass petri dish with a layer of wet towel and filter paper and germinate the seeds in the dark at 25 °C until the optimal root length (2–3 cm).

3. Place the seedlings onto a plastic cover. Group 2–3 seedlings together, thread their roots through the holes in the cover, and position the cover onto a plastic box filled with 2 mM hydroxyurea solution. All roots need to be immersed in the solution. Incubate the roots by aerating in the dark at 25 °C for 18 h.

4. Transfer the cover with the seedlings from the HU solution onto the plastic box filled with 0.1× Hoagland's solution. Incubate the roots by aerating in the dark at 25 °C for 5.5 h.

5. Transfer the cover onto a box containing 2 μM APM solution and incubate the seedlings for 2 h in the dark at 25 °C (without aerating) (*see* **Note 2**).

6. Rinse the roots in a container filled with deionized water (*see* **Note 3**).

7. Collect the roots. Handle the seedlings one by one. Cut off the roots using forceps, remove the excess liquid by touching it to a paper towel, and place the roots in 1.5 mL microcentrifuge tube containing 1 mL of 3:1 fixative. Fix for 1 week in the dark at 37 °C (*see* **Note 4**).

### 3.2   Squash Preparation

1. Stain fixed roots in 2 % carmine acid solution for 2 h (*see* **Note 5**).

2. Cut off the root tip into a drop of 45 % acetic acid on a microscopic slide, remove the root cap (*see* **Note 6**), cover with a glass cover slip, and gently squash the cells by tapping with a

metal needle or a toothpick. Place the edge of a razor blade under the cover slip to spread root cells over a bigger area.

3. Flame the preparation carefully for a few seconds (*see* **Note 7**), press down the cover slip using the ball of your thumb, and then freeze on dry ice for 1 h.

4. Remove the cover slip with a razor blade by lifting the corner edge of the covers slip, soak the slide in 45 % acetic acid for a few seconds at RT, and transfer it into a container with preheated 45 % acetic acid for 3 min at 50 °C.

5. Check and evaluate the quality of dried preparations using light microscopy (*see* **Note 8**). Focus mainly on the number and the quality of metaphase figures and the amount of cytoplasm.

**3.3  Probe Preparation**

General note: In steps that involve working with fluorescently labeled nucleotide(s) and/or probe, keep the tubes in the dark. Work fast and quickly, cover the tubes with aluminum foil or reduce the light intensity in the laboratory, and thus minimize the loss of signal intensity.

1. A set of 15 full-length cDNAs with inserts ranging from 2 to 3.5 kb were used as low-copy probes for FISH (*see* also **Note 9**).

2. Isolate DNA of individual plasmids according to standard alkaline extraction protocols [35].

3. Amplify cDNA sequences using PCR with T3/T7 primers to obtain high-quality probe consisting only of pure cDNA insert (*see* **Note 10**). PCR conditions were 5 min at 94 °C, then 35 cycles of 50 s at 94 °C, 50 s at 55 °C, and 1.5 min at 72 °C. These cycles were followed by 5 min at 72 °C.

4. To produce enough DNA from each cDNA clone, run eight PCR reactions with each clone, mix the PCR products, precipitate DNA by isopropanol precipitation [36], and dissolve in 20 μL TE buffer.

5. Estimate DNA concentration using fluorometer or spectrophotometer (*see* **Note 11**).

6. Label the probes by nick translation. Mix 3 μg DNA, 4 μL 10× nick translation buffer, 4 μL 0.1 M mercaptoethanol, 4 μL non-labeled dNTP's mix, 0.8 μL labeled dUTP (*see* **Note 12**), 4 μL DNA polymerase, 0.4 μL DNase, and ddH₂O up to final volume of 40 μL. Incubate the mix 2 h at 15 °C (*see* **Note 13**).

7. Check the probe quality on the agarose gel (*see* **Notes 14** and **15**).

8. Purify and precipitate cDNA probes. Transfer the probe solution into 1.5 mL microcentrifuge tubes. Add 3 μL of herring sperm DNA, 157 μL of 1× TE buffer, 20 μL of 3 M NaAc, and

500 μL of 96 % chilled ethanol and mix well. After overnight precipitation at −20 °C, centrifuge the probes for 30 min at 4 °C at 14,000 ×*g*, rinse the pellet (*see* **Note 16**) in 70 % ethanol, and dissolve in 10 μL 2× SSC at 37 °C overnight.

***3.4  Slide Post-fixation (See Note 17)***

1. Wash the slides selected in the Subheading 3.2, **step 5**, in 2× SSC for 10 min at RT.

2. Transfer the slides into 45 % acetic acid and wash them for 10 min at RT.

3. Place the slides in 2× SSC for 10 min at RT.

4. Immerse the slide into 4 % formaldehyde and incubate 10 min at RT. **Caution**: Because of formaldehyde toxicity, use the biohazard safety hood in this step.

5. Wash the slides three times in 2× SSC for 5 min at RT.

6. Dehydrate the slides in an ethanol series (70 %, 96 %, and absolute ethanol). 5 min in each ethanol solution. Let the slides dry and use them immediatelly for FISH.

***3.5  Fluorescence In Situ Hybridization***

1. Prepare 20 μL of hybridization mix for each slide by mixing 10 μL of formamide, 5 μL of 4× buffer, 1 μL (~300 ng) of probe (*see* **Note 18**), and ddH$_2$O. Mix well.

2. Pipet the mix onto the slide, cover with a glass cover slip, and glue the edges using rubber cement to prevent evaporation of the mixture.

3. Transfer the slides onto a heating plate and denature the target DNA and probe for 3 min at 80 °C.

4. Place the slides in a humidity chamber at 37 °C overnight. If only directly labeled probes are used in the experiment, continue the protocol with **steps 5–9** of Subheading 3.5. If any indirectly labeled probe is used, skip **steps 5–9** and continue the protocol with **steps 10–26** of Subheading 3.5.

5. Preheat the wash bath and 2× SSC buffer to 57 °C. Transfer the slides into the container with preheated buffer and incubate them for a few min to let the glue moisten.

6. Remove the cover slip and wash the slides in preheated 2× SSC for 20 min at 57 °C.

7. Wash the slide in 2× SSC at RT for 10 min.

8. Dehydrate the preparations by washing in an ethanol series (70, 96, and 100 % ethanol). 5 min in each step.

9. Let the slides dry at RT, and then immediately add Vectashield mounting medium with DAPI and glass cover slip (7 μL per 22 × 22 mm cover slip). Keep the slides away

from light and store them at 4 °C. Continue the protocol from Subheading 3.6.

10. For indirectly labeled probe, preheat the wash bath and 2× SSC and 0.1× SSC buffers at 42 °C. Soak the slides in preheated 2× SSC for a few min to moisten the glue. Then remove the cover slips and wash the slides in 2× SSC for 10 min at 42 °C (*see* **Note 19**).

11. Wash the slides in 0.1× SSC for 5 min at 42 °C.

12. Transfer the slides into 2× SSC buffer and wash them for 10 min at 42 °C.

13. Incubate the slides at RT for 10 min in 2× SSC buffer preheated to 42 °C.

14. Transfer the slides into 4× SSC for 10 min at RT.

15. Remove the slides, let the excess buffer flow off the edge of the slide, and then add 60 μL 1% blocking reagent onto each slide. Cover the preparation with parafilm cut to the size of 22 × 22 mm cover slip, and incubate 10 min at RT.

16. Prepare the mix of blocking reagent and fluorescently labeled antibody to detect the probes indirectly (*see* **Note 20**). The signal from digoxigenin-labeled probe is detected using anti-digoxigenin FITC antibody in a dilution of 1:200 with 1× blocking reagent and biotin-labeled probe using streptavidin Cy3 antibody (1:200).

17. Add 60 μL of antibody with blocking reagent onto each slide, cover it with parafilm cut to cover slip size, and incubate 1 h at 37 °C in humidity chamber.

18. Remove the parafilm and wash the slides three times in preheated 4× SSC at 42 °C.

19. Remove the slides, let the excess buffer flow off the edge of slide, and add 60 μL blocking reagent onto each slide. Cover the preparation with parafilm cut to cover slip size and incubate 10 min at RT.

20. Prepare the mixture of the second layer of antibody in blocking reagent. Dilute anti-sheep FITC (1:1000) and biotinylated anti-streptavidin (1:1000) in blocking reagent.

21. Add 60 μL of the mix onto each slide, cover with parafilm cut to cover slip size, and incubate 1 h in humidity chamber at 37 °C.

22. Repeat **steps 18** and **19** of Subheading 3.5.

23. Prepare the third layer of detection for biotin-labeled probe. Add streptavidin Cy3 into blocking reagent in dilution 1:200.

24. Pipet the mixture on the slide, cover it with parafilm cut to cover slip size, and incubate another 1 h at 37 °C in humidity chamber.

25. Remove the parafilm and wash the slides three times in preheated 4× SSC at 42 °C and then dry the slides at RT.

26. Add Vectashield mounting medium with DAPI and apply a cover slip (7 μL per 22×22 mm cover slip). Keep the slides away from light and store them at 4 °C.

### 3.6 Microscopy (See Note 21)

1. Signal capture: When scanning the slide with the fluorescent microscope, select fluorescence filter blocks according to the expected signal intensity. In order to prevent fading of weak signals, capture first the signal of the probe giving the weakest signal, then the stronger one, and then finally the DAPI counterstaining. This protects the photons from the short probe and allows them to excite at once as a signal of maximal intensity.

2. Hybridization efficiency (*see* **Note 22**): Capture images of as many metaphase figures as possible. Select the cell plates with your chromosomes of interest having no other chromosomes or cellular debris overlapping to increase the probability of clearly observing the hybridization signal. It is not uncommon to observe the hybridization signal only on one of the homologues (*see* **Note 23**).

3. Signal intensity (*see* **Note 24**): The fluorescent signal with cDNA probes is very weak. Once a perfect figure is identified, do not attempt to observe fluorescence signals, but focus the image rapidly and capture the fluorescence signal before photo bleaching of the signal. Avoid shiny spots in the visual field, which may interfere with the adjustment of fluorescence capture time of camera.

## 4  Notes

1. In general, the localization of short probes using FISH results in lower hybridization efficiency. The hybridization signal is observed only in about 40 % of examined metaphase plates (*see* ref. 37). Thus, it is important to use high-quality preparations with sufficient number of cells in metaphases.

2. Alternatively, the frequency of metaphase cells can be increased by cold water treatment (*see* ref. 38). It is an easy method that does not involve special protocols and additional laboratory equipment. However, the number of dividing cells is considerably lower.

3. If the roots cannot be collected immediately after the APM treatment, the holder with seedlings should be placed in a container with ice water containing ice cubes and kept overnight in a refrigerator.

4. The process of fixation may differ among plant species. The described timing and fixative composition are optimal for barley.

5. Stained roots can be stored in fixative at −20 °C. Optional: place the stained roots into carmine solution for 10 min before squashing.

6. Optional: soak the microscopic slides in 96 % ethanol overnight to wash out the grease and other dirt. Do not touch the clean slides with bare fingers.

7. Heating the preparation is one of the critical steps in the protocol. The heating reduces the amount of cytoplasm and leads to cytoplasm-free preparation, which is essential for detecting weak hybridization signals. Insufficient flaming leaves a considerable amount of residual cytoplasm. On the other hand, overheating burns the cells and irreversibly damages the chromatin structure. Flame the slides up to the boiling point. Let all acetic acid evaporate and stop the heating immediately when air bubbles appear under the cover slip.

8. Preparations from stained root tips can be observed using standard objectives. In order to observe unstained tissues, it is recommended to use phase contrast microscopy.

9. Fl-cDNAs can be obtained from a collection of 5006 full-length cDNA sequences of barley (*see* ref. 39) or similar sources from the plant species of choice. The fl-cDNAs were cloned in pFLCIII-sfi-cDNA in *Xho*i/*Sal*I and *Bam*HI cloning site.

10. If the cloning primers are not available, whole plasmids can be used as probes for FISH. However, the plasmid sequence represents the majority of the sequence in the labeled probe, which may lead to nonspecific signals and decrease the signal-to-noise ratio.

11. Because of the high amount of DNA needed for nick translation reaction, DNA concentration has to exceed 150 ng/μL. If the concentration is lower, it is recommended to repeat the amplification and merge more PCR products until the DNA amount is sufficient.

12. FISH with directly labeled probes is faster, results in lower background signals, and provides the opportunity to use more than two differentially labeled probes in one experiment. Unfortunately, there are not enough fluorochromes with sufficient quantum yield (except for Texas Red and perhaps also

Alexa 488) to give detectable hybridization signals with short probes. FISH with indirectly labeled probes results in much brighter signals; however, the signal-to-noise ratio is inferior as compared to FISH with directly labeled probes.

13. There are numerous nick translation labeling kits available on the market. Their use is convenient and simplifies the protocol. However, preparation of the reaction mix provides the opportunity to adjust reaction conditions for each particular experiment (*see* ref. 40).

14. The probe size should range from 100 to 500 bp. Increase the amount of DNase in the reaction if the fragments are too long. Do not prolong the reaction time over 2 h. The warranted enzyme lifetime is around 2 h, after which its efficiency declines.

15. Optional: if the probe is labeled directly, it is possible to verify the amount of incorporated fluorochrome as follows. Take an image of the electrophoretically separated probe before and after staining in ethidium bromide and compare the pictures. The probe fragments of the appropriate length should be visible on the gel without EtBr staining.

16. If a directly labeled probe is used, the pellet should have the color of the label, e.g., the pellet of Texas Red-labeled probe should be purple. If this is not the case, the fluorochrome was not incorporated into the probe and remained in the solution.

17. Postfix and dehydrate the slides just before use. During the storage, the dehydrated slides reabsorb water from the air and will negatively affect the chromosome shape and signal structure after hybridization.

18. Optional: it is recommended to use a previously verified probe in combination with a single-copy sequence in the same experiment. The verified probe will serve as a control of the FISH procedure. If possible, the marker should not co-localize with unknown probe (Fig. 1a).

19. If the water bath is equipped with a shaker, all washes can be done by shaking. Gentle shaking improves the washing efficiency.

20. Fluorescent signal of short probes is very weak. Therefore, the detection of hybridization signals after FISH with indirectly labeled probes requires signal amplification to increase the final signal intensity.

21. When observing hybridization signals after FISH with very short probes, one must be precise, fast, and efficient. There is usually only just one chance to take a good picture for publication of outcomes, as the hybridization signal bleaches quickly.

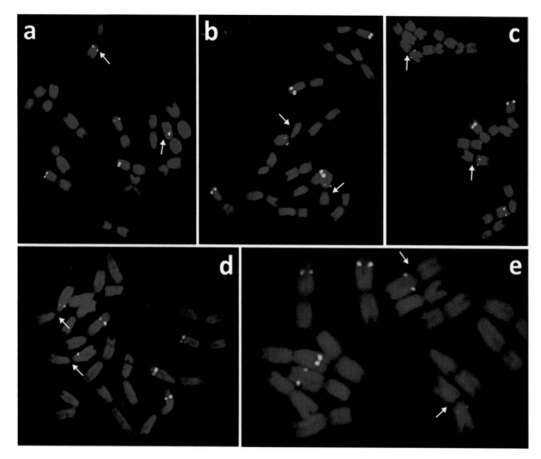

**Fig. 1** Simultaneous localization of fl-cDNA clones (*purple*) and 5S rDNA (*green*) on barley mitotic metaphase chromosomes. Probes for fl-cDNA clones FLbaf140k15 (**a**), FLbaf54a18 (**b**), and FLbaf169o18 (**c–e**) were directly labeled by *Texas Red*. The probe for 5S rDNA was directly labeled by fluorescein isothiocyanate (FITC). The *arrows* indicate the position of cDNA clone hybridization signals on the short (**a**) and long (**b–e**) arms of chromosome 7H. Note differences in the signal intensity for clone FLbaf169o18: (**c**) double signals on sister chromatids of both homologues, (**d**) signal on only one sister chromatid of both homologues, and (**e**) combination of both hybridization patterns. The chromosomes were counterstained by DAPI (*blue*)

After capturing the signal once, it becomes very weak, making it virtually impossible to acquire another sharp image again.

22. The protocol is highly reliable. The hybridization signal is observed in more than 90% of examined figures, which is nearly three times more frequent than reported for other protocols (*see* ref. 39). Nevertheless, the success of the hybridization strongly depends on local chromatin structure. Therefore, sometimes a smaller probe can result in better hybridization signals than a longer one.

23. Hybridization efficiency of short probes can vary among the figures on one slide and even between the homologous chromosomes in one metaphase plate. The probe hybridization

strongly relies on local chromatin structure, and differences in chromatin conformation and/or local chromatin damage may result in variation of the signal appearance. Generally, hybridization signal is detected on both chromatids as double dots (Fig. 1c, e), but it is not unusual that the low-copy signal is observed only on one of the sister chromatids (Fig. 1d, e).

24. Fluorescent signals can be very tiny (Fig. 1b), and signals from probes shorter than 5 kb are usually invisible to human eye and must be captured by the microscope camera and observed on a computer screen. The interpretation of FISH results and their analysis puts high demands on scientist's skills and requires long-term experience. Apart from the personal factor, the success is strongly dependent on other parameters such as high-quality cytological preparations, specific and concentrated probe labeled with a bright fluorochrome, and, last but not the least, high-quality fluorescence microscope equipped with a sensitive, low-noise CCD camera.

## Acknowledgment

We thank Zdeňka Dubská for technical assistance with cell cycle synchronization. Barley fl-cDNA clones were kindly provided by Kazuhiro Sato (Okayama University, Japan). This work was supported by a grant award LO1204 from the National Program of Sustainability I.

## References

1. Dear PH (2001) Genome mapping. In: eLS. John Wiley & Sons Ltd, Chichester. http://www.els.net. Nature Publishing Group, London, UK, pp 1–7, www.els.net

2. The Arabidopsis Genome Initiative (2000) Analysis of the genome sequence of the flowering plant *Arabidopsis thaliana*. Nature 408:796–815

3. Goff SA, Ricke D, Lan TH et al (2002) A draft sequence of the rice genome (*Oryza sativa* L. ssp. *japonica*). Science 296(5565):92–100

4. Schnable P, Ware D, Fulton RS et al (2009) The B73 maize genome: complexity, diversity, and dynamics. Science 326:1112–1115

5. Gordo SM, Pinheiro DG, Moreira EC et al (2012) High-throughput sequencing of black pepper root transcriptome. BMC Plant Biol 12:168

6. Varshney RK, Kudapa H, Roorkiwal M et al (2012) Advances in genetics and molecular breeding of three legume crops of semi-arid tropics using next-generation sequencing and high-throughput genotyping technologies. J Biosci 37:811–820

7. Faino L, Thomma BP (2014) Get your high-quality low-cost genome sequence. Trends Plant Sci 19:288–291

8. Karafiátová M, Bartoš J, Kopecký D et al (2013) Mapping nonrecombining regions in barley using multicolor FISH. Chromosome Res 21:739–751

9. Lou Q, Zhang Y, He Y et al (2014) Single-copy gene-based chromosome painting in cucumber and its application for chromosome rearrangement analysis in *Cucumis*. Plant J 78:169–179

10. Poursarebani N, Ma L, Schmutzer T et al (2014) FISH mapping for physical map improvement in the large genome of barley: a case study on chromosome 2H. Cytogenet Genome Res 143:275–279

11. Shearer LA, Anderson LK, de Jong H et al (2014) Fluorescence *in situ* hybridization and optical mapping to correct scaffold arrangement in the tomato genome. G3 30:1395–1405

12. Trask B, Trask B (1999) Fluorescence *In situ* Hybridization. In: Birren B, Green ED, Hidžer P, Klapholz S, Myers RM, Riethman H, Roskams J, Birren B, Green ED, Hidžer P, Klapholz S, Myers RM, Riethman H, Roskams J (eds) Genome analysis (a laboratory manual)-Mapping genome, vol 4. Cold Spring Harbor Laboratory Press, USA, pp 303–407

13. Cheng Z, Buell CR, Wing RA et al (2002) Resolution of fluorescence *in situ* hybridization mapping on rice mitotic prometaphase chromosomes, meiotic pachytene chromosomes and extended DNA fibers. Chromosome Res 10:379–387

14. de Jong JH, Fransz P, Zabel P (1999) High-resolution FISH in plants-techniques and applications. Trends Plant Sci 4:258–263

15. Schubert V, Meister A, Tsujimoto H et al (2011) Similar rye A and B chromosome organization in meristematic and differentiated interphase nuclei. Chromosome Res 19:645–655

16. Valárik M, Bartoš J, Kovářová P et al (2004) High resolution FISH on super-stretched flow-sorted chromosomes. Plant J 27:940–950

17. Jiang J, Gill BS (2006) Current status and the future of fluorescence in situ hybridization (FISH) in plant genome research. Genome 49:1057–1068

18. Künzel G, Korzun L, Meister A (2000) Cytologically integrated physical restriction fragment length polymorphism maps for the barley genome based on translocation breakpoints. Genetics 154:397–412

19. Jiang J, Gill B, Wang G-L et al (1995) Metaphase and interphase fluorescence *in situ* hybridization mapping of the rice genome with bacterial artificial chromosomes. Proc Natl Acad Sci 92:4487–4491

20. Szakács E, Molnár-Láng M (2010) Identification of new winter wheat - winter barley addition lines (6HS and 7H) using fluorescence *in situ* hybridization and the stability of the whole 'Martonvásári 9 kr1' - 'Igri' addition set. Genome 53:35–44

21. Danilova TV, Friebe B, Gill BS (2012) Single-copy gene fluorescence *in situ* hybridization and genome analysis: Acc-2 loci mark evolutionary chromosomal rearrangements in wheat. Chromosoma 121:597–611

22. Lysák M, Fransz PF, Ali BH et al (2001) Chromosome painting in *Arabidopsis thaliana*. Plant J 28:689–697

23. Gan Y, Chen D, Liu F et al (2011) Individual chromosome assignment and chromosomal collinearity in *Gossypium thurberi*, *G trilobum* and D subgenome of *G. barbadense* revealed by BAC-FISH. Genes Genet Syst 86:165–174

24. Cheng Z, Presting GG, Buell CR et al (2001) High-resolution pachytene chromosome mapping of bacterial artificial chromosomes anchored by genetic markers reveals the centromere location and the distribution of genetic recombination along chromosomes 10 of rice. Genetics 157:1749–1757

25. Wolny E, Fidyk W, Hasterok R (2013) Karyotyping of *Brachypodium pinnatum* (2n = 18) chromosomes using cross-species BAC-FISH. Genome 56:239–243

26. Zhang P, Li W, Friebe B et al (2004) Simultaneous painting of three genomes in hexaploid wheat by BAC-FISH. Genome 47:979–987

27. Suzuki G, Ogaki Y, Hokimoto N et al (2011) Random BAC FISH of monocot plants reveals differential distribution of repetitive DNA elements in small and large chromosome species. Plant Cell Rep 31:621–628

28. Janda J, Šafář J, Kubaláková M et al (2006) Advanced resources for plant genomics: BAC library specific for the short arm of wheat chromosome 1B. Plant J 47:977–986

29. Schnabel E, Kulikova O, Penmetsa RV et al (2003) An integrated physical, genetic and cytogenetic map around the sunn locus of *Medicago truncatula*. Genome 46:665–672

30. Tang X, de Boer JM, van Eck HJ et al (2009) Assignment of genetic linkage maps to diploid *Solanum tuberosum* pachytene chromosomes by BAC-FISH technology. Chromosome Res 17:399–415

31. Danilova TV, Friebe B, Gill BS (2014) Development of a wheat single gene FISH map for analyzing homoeologous relationship and chromosomal rearrangements within the Triticeae. Theor Appl Genet 127:715–730

32. Fuchs J, Schubert I (1995) Localization of seed protein genes on metaphase chromosomes of *Vicia faba* via fluorescence *in situ* hybridization. Chromosome Res 3:94–100

33. Yang K, Zhang H, Converse R et al (2011) Fluorescence *in situ* hybridization on plant extended chromatin DNA fibers for single-copy and repetitive DNA sequences. Plant Cell Rep 30:1779–1786

34. Kato A, Vega JM, Han FP et al (2005) Advances in plant chromosome identification and cytogenetic techniques. Curr Opin Plant Biol 8:148–154

35. Birnbiom HC (1983) A rapid alkaline extraction method for the isolation of plasmid DNA. Methods Enzymol 100:243–255

36. Fischer JA, Favreau MB (1991) Plasmid purification by phenol extraction from guanidinium thiocyanate solution: development of an automated protocol. Anal Biochem 194:309–315

37. Schwarzacher T, Heslop-Harrison P (2000) Practical *in situ* hybridization. BIOS Scientific Publishers Limited, Oxford

38. Mirzaghaderi G (2009) A simple metaphase chromosome preparation from meristematic root tips cell of wheat for karyotyping or *in situ* hybridization. Afr J Biotechnol 9:314–318

39. Sato K, Shin-I T, Seki M et al (2009) Development of 5006 full-length cDNAs in barley: a tool for accessing cereal genomics resources. DNA Res 16:81–89

40. Kato A, Albert PS, Vega JM et al (2006) Sensitive fluorescence *in situ* hybridization signal detection in maize using directly labeled probes produced by high concentration DNA polymerase nick translation. Biotech Histochem 81:71–78

# Chapter 6

# Immunolabeling and In Situ Labeling of Isolated Plant Interphase Nuclei

## Ali Pendle and Peter Shaw

## Abstract

Specific labeling of proteins and nucleic acids by immunofluorescence or in situ techniques is an important adjunct to microscopical analysis for cell biology. Labeling of nuclear structures in intact complex tissues is often hampered by problems of penetration of the macromolecular labeling reagents needed. Here we describe a method of labeling isolated plant nuclei that we have found to be a useful approach that can help to overcome these problems.

**Key words** Immunolabeling, In situ hybridization, Plant cell biology, Nuclei, Nuclei isolation, Fluorescence microscopy

## 1 Introduction

Imaging of interphase nuclei by optical microscopy methods such as phase or differential interference contrast, or with fluorescence microscopy using a general DNA dye like DAPI, shows the overall shape and some substructural features, as, for example, nucleoli and heterochromatin. However, detailed imaging requires labeling of specific proteins or other components. Ideally, when analyzing living systems, it is best to image live organisms and cells by expressing proteins fused to tags such as GFP. But this is not always possible. Alternative methods are antibody labeling (immunofluorescence) or in situ hybridization to RNA or DNA sequences [1, 2] followed by fluorescent detection. These procedures require access of large molecules to the nuclear interior, which in turn requires opening up the structure to enable diffusion of these molecules; the resulting techniques are always a balance between efficiency of labeling and structural preservation of the specimens under investigation.

In multicellular plants, nuclei are located inside cells, which are surrounded with relatively impervious cell walls, and are embedded within tissues containing many cells and often multiple layers

Shahryar F. Kianian and Penny M.A. Kianian (eds.), *Plant Cytogenetics: Methods and Protocols*,
Methods in Molecular Biology, vol. 1429, DOI 10.1007/978-1-4939-3622-9_6,
© Springer Science+Business Media New York 2016

of cells. This can make both the penetration of labeling reagents and the subsequent microscopy imaging either challenging or impossible. Often it is necessary to section the material before labeling (e.g., [3, 4]), or gently squashing it onto the slide. A very useful alternative is to image isolated nuclei. Here we describe methods for isolating plant nuclei as preparations on microscope slides or coverslips using a cytospin centrifuge and labeling them by immunofluorescence or in situ hybridization. We have described these procedures for *Arabidopsis thaliana* seedling roots, but very similar methods can be used with minimal modification for other parts of the plant and for other species.

Two methods for releasing nuclei from roots are described below. Alternatively the simplest method is to repeatedly chop the roots with a sharp razor blade (not described). This is surprisingly effective for the small amounts needed for microscopy labeling. Once a technique for nuclei isolation has been established, the most common reason for failure is poor fixation. Formaldehyde is usually used for fixation of plant material for light microscopy. We advise that formaldehyde be freshly made from paraformaldehyde as described below, as it degrades in solution. Electron microscopy requires better fixation than formaldehyde can provide and glutaraldehyde is usually the fixative of choice. Small amounts of glutaraldehyde are sometimes added to formaldehyde for light microscopy, which can improve preservation, particularly for harsh treatments such as in situ labeling. But often the better preservation of cells prevents diffusion of the labeling probes into the specimen. Glutaraldehyde also causes a large degree of background fluorescence. This can be alleviated to some extent by treatment with sodium borohydride.

Some of the most informative studies of the nucleus involve the use of both immunofluorescence and in situ labeling on the same specimen, for example, to show the association of particular proteins with specific genes, or other DNA or RNA sequences. In these cases, it is generally best to carry out at least the primary antibody labeling before the in situ. This is presumably because the harsh denaturation conditions for in situ labeling destroy the antigenicity of the proteins, whereas the complexed antibody-antigen is extremely stable. However, each scenario is different and needs careful monitoring to determine the best sequence of operations. An example of a nuclear preparation labeled by two different immunofluorescence probes is shown in Fig. 1.

## 2    Materials

### 2.1    Plant Material Preparation

1. Sterile 9 cm square Petri dishes for plant growth media.

2. Plant growth medium: Murashige and Skoog (M&S) [0.025 mg/l $CoCl_2 \cdot 6H_2O$, 0.025 mg/l $CuSO_4 \cdot 5H_2O$, 36.7 mg/l Na Fe-EDTA, 6.2 mg/l $H_3BO_3$, 0.83 mg/l KI, 16.9 mg/l $MnSO_4 \cdot 2H_2O$, 0.25 mg/l $Na_2MoO_4 \cdot 2H_2O$,

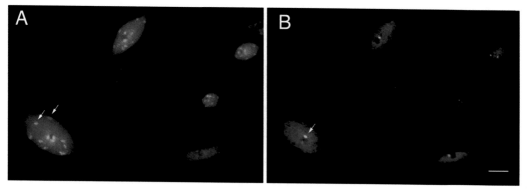

**Fig. 1** *Arabidopsis* labeled nuclei with two different immunofluorescence. Nuclei from *Arabidopsis* roots were prepared as described and labeled by immunofluorescence using (**a**) DAPI (*blue*) and (**b**) antibody 4G3 (*green*). Cells of different sizes and thus different ploidy levels are seen in the field of view. The bright DAPI foci are regions of centromeric heterochromatin, corresponding to the ten chromosomes (examples of two chromosomes identified with *arrows* in **a**). In polyploid cells (larger size cells), substructure is often seen in the heterochromatin foci, suggesting association of multiple centromeres. 4G3 labels the spliceosomal protein U2B and in plants shows strong labeling of Cajal bodies (example is identified with *arrows* in **b**) (*see* ref. 5)

8.6 mg/l $ZnSO_4 \cdot 7H_2O$, 332.02 mg/l $CaCl_2 \cdot 2H_2O$, 170 mg/l $KH_2PO_4$, 1900 mg/l $KNO_3$, 180.5 mg/l $MgSO_4 \cdot 7H_2O$, 1650 mg/l $NH_4 \cdot NO_3$, pH 5.8].

3. 10 % v/v bleach (store brands contains 5–10 % sodium hypochlorite) solution in $dH_2O$.

4. *Arabidopsis* seeds.

**2.2 Cytofunnel Preparation**

1. Microscope slides with frosted end.

2. Shandon Cytospin 4 Cytocentrifuge (Thermo Scientific).

3. Shandon Single White Cytofunnels (Thermo Scientific).

4. Shandon Cytoclips™ (Thermo Scientific).

**2.3 Immunofluorescence**

1. Triton TX-100. Prepare a stock of 10 % v/v Triton TX-100 in $dH_2O$ and store at 4 °C.

2. Dilute sulphuric acid. Prepare a solution of 10 % v/v sulphuric acid by the careful drop-wise addition of concentrated (98 %) sulphuric acid to $dH_2O$.

3. The pH 4.5–10 indicator strips.

4. Phosphate buffer saline (PBS) pH 7.0 (for medium). 10× PBS stock (10× PBS: 1.37 M NaCl, 27 mM KCl, 100 mM $Na_2HPO_4$, 18 mM $KH_2PO_4$).

5. Vacuum infiltration equipment. A plastic vacuum dessicator is attached to a rotary vacuum pump. This equipment should be situated in a fume hood.

6. Nuclei isolation buffer (NIB): 10 mM MES (2-(N-morpholino) ethanesulfonic acid) pH 5.5, 0.2 mM sucrose, 2.5 mM EDTA,

2.5 mM DTT, 10 mM NaCl, 10 mM KCl, 0.1 mM Spermine, 0.5 mM spermidine, 0.5 % Triton TX-100.

7. Flat-ended stainless steel rod (140 mm×3 mm) and/or stainless steel grinder for 1.5 ml microcentrifuge tube (*see* **Note 1** and Fig. 2).

8. Nylon mesh filter – either CellTrics disposable 30 μm filter (Partec) or homemade (*see* **Note 2** and Fig. 3).

9. Blocking solution: 3% w/v bovine serum albumin (BSA) in PBS pH 7.0. Make fresh each time.

10. Homemade plastic coverslips made from transparent autoclave bags cut into the standard coverslip size (22 mm by 22 mm).

11. 4′, 6-Diamidino-2-phenylindole (DAPI) 1 μg/ml solution in dH₂O. Protect from light and store at 4 °C.

12. 2, 2′-Thiodiethanol (TDE). 97% v/v TDE, 3% v/v PBS, pH 7.0. Store at 4 °C and protect from light.

13. Vectashield anti-fade mounting, Vector Laboratories.

14. Coverslips. Carl Zeiss high-performance coverslips No 1.5.

15. Nail varnish/nail polish.

16. Glass embryo dishes (30 mm).

**Fig. 2** Equipment for maceration as described in *see* **Note 1**. (**a**) Stainless steel rod with flat end. (**b**) Stainless steel grinder made to the internal profile of an Eppendorf tube (**c**)

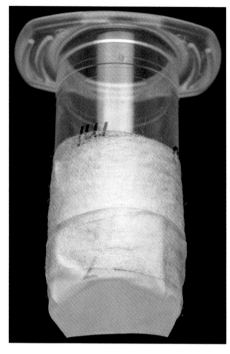

**Fig. 3** Homemade 30 μm filter assembly (*see* **Note 2**)

***2.4 In Situ Hybridization***

1. Nucleic acid probes labeled with digoxigenin or biotin (*see* **Note 3**).

2. Formamide, deionized, minimum 99.5 % (*see* **Note 4**).

3. Formamide, laboratory reagent grade.

4. Dextran sulfate.

5. 20× SSC (saline sodium citrate buffer). 3 M NaCl, 300 mM Tri-sodium citrate ($Na_3C_6H_5O_7$) pH 7.0.

6. SDS (sodium dodecyl sulfate), preferably purchased as a 20 % solution in $dH_2O$, to avoid handling the solid powder.

7. Salmon sperm DNA.

8. OmniSlide hybridization chamber (*see* **Note 5**).

9. Tween 20.

10. Ethanol.

# 3  Methods

Carry out all procedures at room temperature unless otherwise stated.

***3.1 Preparation of Plant Material***

1. Prepare M&S media plates. Use Murashige and Skoog medium supplemented with 0.5 % (w/v) Phytagel and 1 % (w/v) sucrose; for 1 l of solution, autoclave for 20 min at 120 °C;

allow to cool to about 60 °C before pouring into 10 cm square Petri dishes while still molten under sterile conditions (*see* **Note 6**). Use approximately 60 ml solution per dish. Allow to cool and solidify before use.

2. Surface sterilize *Arabidopsis thaliana* seeds in 10% bleach for 10 min in 1.5 microcentrifuge tube, then wash with three changes of sterile water.

3. Plate out individual seeds at 2–3 mm spacing in two rows across prepared Petri dishes with M&S media, allowing space for root growth.

4. Stratify the seeds by incubating for 2 days at 4 °C in the dark (*see* **Note 7**).

5. Germinate and grow seedlings by placing plates vertically (*see* **Note 8**) in a 25 °C growth chamber under constant illumination (*see* **Note 9**). Use approximately 5-day-old seedlings for the preparation of nuclei. Older plants can be used to enrich for endoreduplicated nuclei, whereas younger plants can be used to enrich for diploid nuclei (such as in meristematic cells). Also *see* **Note 10** on glasshouse-grown plants.

*3.2 Preparation of Fixative*

Prepare an 8% w/v solution using prilled paraformaldehyde (*see* **Note 11**). Make the solution by adding paraformaldehyde to dH$_2$O on a heated stirrer in a fume cupboard (*see* **Note 11**). Warm to approximately 60 °C and make alkaline by the addition of a few drops of 1 M NaOH. The paraformaldehyde should dissolve to give a clear solution of 8% formaldehyde. Immediately prepare a solution of 4% formaldehyde in PBS by adding an equal volume of 2× PBS pH 7.0 to the 8% formaldehyde solution. This will give a final concentration of 4% formaldehyde in 1× PBS (*see* **Note 11**). Adjust the pH to 7 using dilute H$_2$SO$_4$ (*see* **Note 12**). Add Triton TX-100 to 0.01%.

*3.3 Assembly of Cytofunnel Unit*

1. Place a plain glass slide with frosted end into the cytoclip, keeping the frosted end to the outside of the clip.

2. Position a single white cytofunnel over the slide and secure with the cytoclip. Label appropriately. Figure 4 shows the cytofunnel components separately (Fig. 4a, b, and c) and assembled ready to load into the cytospin centrifuge (Fig. 4d).

*3.4 Immuno- fluorescence Procedure*

1. Cut root tips (up to 10 mm in length) (*see* **Note 13**) from *Arabidopsis* seedlings while still on plates. Collect 50–100 root tips and place into 20 ml of fixative in a 30 ml glass bottle.

2. Vacuum infiltration of fixative. Place the open glass bottle containing the fixative and tissue samples in the vacuum dessicator and replace the dessicator lid. Switch on the vacuum pump and

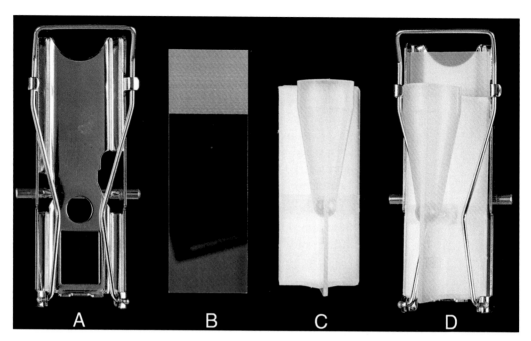

**Fig. 4** Cytofunnel components and assembly. (**a**) Cytofunnel assembly clip. (**b**) Microscope slide with frosted end. (**c**) Plastic cytofunnel. (**d**) Assembled apparatus (*see* Subheadings 2.2 and 3.3)

open the vacuum valve carefully to slowly pump out the air until the fixative solution bubbles gently. After about 5 min, release the vacuum and see if the tissue sinks in the fixative solution. If it still floats, repeat the vacuum procedure.

3. Incubate in the fixative for 1 h.

4. Wash roots in PBS pH 7.0 for 10 min, repeat twice.

5. Place washed roots into 300–400 μl of nuclei isolation buffer (NIB) in a 30 mm glass embryo dish and macerate roots vigorously with a flat-ended stainless steel rod (*see* **Note 14**). Continue macerating for several minutes until the roots have been reduced to tiny pieces releasing the nuclei into the NIB. Alternatively nuclei can be extracted by placing the fixed roots into a 1.5 ml microcentrifuge tube with the NIB, and then a stainless steel grinder can be used to grind the material to release the nuclei.

6. Filter the nuclei solution through a 30 μm nylon mesh filter (*see* **Note 15**).

7. Pipette 50 μl of the filtered nuclei into each assembled cytofunnel and spin in the cytospin at 500 rpm ($30 \times g$) for 3 min.

8. Disassemble the cytofunnel units, remove the slides, and allow them to air-dry for 40–50 min.

9. Immerse the slides in 70% ethanol for 30 min.

10. Wash with PBS, pH 7.0 for 10 min, repeat twice.

11. Block tissue with 3% BSA in PBS pH 7.0 for 1 h (*see* **Note 16**).

12. Apply primary antibodies diluted appropriately in blocking solution (3% BSA in PBS) and incubate for a minimum of 2 h at room temperature, or up to a maximum of overnight at 4 °C. It is important to avoid drying of the solutions on the slides; place incubating slides in a sealed container such as a plastic Petri dish along with moist filter paper and/or use plastic autoclave bag coverslips.

13. Wash with PBS, pH 7.0 for 10 min, repeat five times.

14. Apply appropriate secondary antibodies diluted in blocking solution and incubate for 2 h at room temperature.

15. Wash with PBS, pH 7.0 for 10 min, repeat five times.

16. Counterstain for DNA with a 1 μg/ml solution of DAPI in $H_2O$ for 30 min.

17. Wash with PBS, for 10 min, repeat once.

18. Remove as much liquid as possible and add 10–15 μl of a suitable mounting medium (*see* **Note 17**) and cover with a glass cover slip (*see* **Note 18**).

19. Seal the coverslip to the slide with nail varnish.

20. View samples with a suitable microscope.

*3.5 In Situ Hybridization Procedure*

1. Follow procedure for immunofluorescence to step **9** of Subheading 3.4.

2. Immerse slides in 100% ethanol for 10 min and then allow to air-dry.

3. Apply 25 μl of a hybridization mixture (7 ng/μl labeled DNA, 50% formamide, 10% dextran sulfate, 2× SSC, 0.125% SDS, 1 μg/μl salmon sperm DNA) to the slides and cover with a plastic coverslip.

4. Denature the probes and tissue simultaneously at 75 °C for 10 min and allow hybridization to proceed at 37 °C in a humid chamber for at least 16 h (*see* **Note 5**).

5. Following hybridization, remove the coverslips and wash the slides for 5 min successively in 2× SSC at 42 °C, then twice in 20% formamide, 0.1× SSC at 42 °C, twice in 2× SSC at 42 °C, twice in 2× SSC at room temperature, twice in 4× SSC/0.2% Tween 20 at room temperature, and finally in PBS pH 7.0.

6. Incubate the slides in blocking solution for 1 h.

7. Apply an appropriate fluorescently labeled antibody for digoxigenin or fluorescent labeled avidin/streptavidin/extravidin for

biotin, diluted in blocking solution, to the slides, and incubate for 2 h.

8. Continue from **step 13** of the immunofluorescence procedure of Subheading 3.4 to end.

**3.6 Combined Immunofluorescence and In Situ Hybridization Procedure**

1. Follow the Immunofluorescence procedure of Subheading 3.4 to **step 13**.

2. Fix in 4% paraformaldehyde for 10 min at room temperature (optional) (*see* **Note 19**).

3. Wash in PBS, pH 7.0 for 10 min, repeat twice.

4. Follow the in situ hybridization procedure of Subheading 3.5 from **steps 3–6**.

5. Apply fluorescently labeled secondary antibodies to detect the probe labels (digoxigenin or biotin) and to recognize the primary antibodies used for the immunofluorescence. The secondary antibodies should be applied diluted in blocking buffer and incubated for 2 h.

6. Follow **steps 15–20** of the immunofluorescence procedure of Subheading 3.4.

# 4  Notes

1. The stainless steel maceration rod was made by cutting a 140 mm length of 3 mm diameter rod and removing any sharp edges by gently grinding the cut edges to leave a flat-bottomed rod (Fig. 2a). The grinder fitting 1.5 microcentrifuge tube was turned from a 20 mm × 8 mm piece of stainless steel rod to give the internal shape of an Eppendorf tube (10° angle) with the end rounded to fit the bottom of the tube. This was screw tapped to accept the 5 mm stem also made from stainless steel that is screwed into the head. A plastic handle can be added for comfort (Fig. 2b, c).

2. A homemade filter can be made from the body of a 20 ml syringe with the tip cutoff. The cut end is then covered with a piece of 30 μm nylon net filter (Millipore) and secured with tape (Fig. 3).

3. Make nucleic acid probes using standard techniques to incorporate tags, which can subsequently be detected by relevant antibodies [1, 2]. The technique described in this protocol is for detection of DNA probes but a similar procedure can be followed for the detection of RNA probes.

4. A high purity grade of formamide should be used in the hybridization mixture, but laboratory reagent grade is sufficient for the washing solutions.

5. We used a Hybaid OmniSlide Thermal Cycler for the hybridization step. This has a built in humidity chamber that prevents drying out of the hybridization mixture on the slides. We used a program with an initial denaturation step of 75 °C followed by a gradual ramp down of the temperature to a final hybridization temperature of 37 °C, which was then held for at least 16 h. Unfortunately, this Thermal Cycler is no longer commercially available, although other models are available. However, a humidity chamber can easily be made from a plastic box with a fitted lid lined with wet tissue paper and a sheet of plastic to lay the slides on. Slides can be denatured on a flat bed heat block, then quickly transferred to the humid chamber and incubated at 37 °C.

6. It is not possible to remelt Phytagel, so allow the medium to cool to a reasonable working temperature of about 60 °C before pouring the plates under sterile conditions.

7. Stratification of seeds by a cold treatment of 2 days will ensure an even germination rate.

8. By placing the plates vertically, the germinating roots will grow along the surface of the gel and can be removed easily without damage to the root structure.

9. We use 25 °C, constant light growth conditions to germinate *Arabidopsis* seedlings. However, this procedure works equally well with seedlings grown at cooler temperatures with light/dark cycles. It should be noted that different conditions need different times to reach the same stage of development.

10. Although healthy looking plants can be grown year round in today's glasshouses and controlled environment rooms, it is a frequent observation, although we have never seen documented, that the best immunolabeling or in situ labeling is obtained with plants grown close to their native growing season. For example, wheat in the UK gives the best labeling results when grown between April and October, even in glasshouses. In southern Spain, wheat produces the best results in the winter and very poor results in the hot summers. We are unsure of the reason for this; it is likely that even the best controlled environment rooms cannot completely remove the influence of the external environment, and this stresses the plants to some extent, even though they look healthy. This in turn could influence the production and composition of such components as the cell walls, which may make penetration of labeling reagents more difficult.

11. We find it best to make fresh formaldehyde each time. Formaldehyde may cause cancer so should always be used in a fume hood. Weigh out paraformaldehyde in the fume hood, wearing appropriate safety clothing, lab coat, gloves, and eye

protection. Warm the solution but do not allow it to boil, as this will degrade the formaldehyde. We recommend using paraformaldehyde "prilled" rather than powder to avoid harmful dust. Paraformaldehyde dissolves best at alkaline pH. Therefore it is best to make it in dH$_2$O with a few drops of alkali (rather than in a buffer) and then add a 2× buffer once the formaldehyde is dissolved. In this manner, less acid is needed to bring the pH back to neutral. If more than a ten drops of alkali per 100 ml of liquid are needed to dissolve the paraformaldehyde, this is probably a sign that the paraformaldehyde has degraded and should be replaced. Paraformaldehyde should be kept dry at all times. It lasts longer when stored at 4 °C in the dark than at room temperature, but should be allowed to warm to room temperature before opening the container to avoid condensation.

12. To adjust the pH of formaldehyde, do NOT use HCl, as reaction of formaldehyde and HCl produces the carcinogen Bis (chloromethyl) ether. Use pH strips to determine the pH rather than a pH electrode, as fixatives can degrade pH electrodes.

13. Using the first 10 mm of the root tip will ensure the availability of nuclei from both meristematic and differentiated tissue.

14. The maceration step can take several minutes of continuous stabbing with the flattened tip of the stainless steel rod to effectively release a substantial amount of nuclei. A good guide is to reach a point when there are very small pieces of root remaining and the solution is partially cloudy.

15. The nylon mesh filter should be wetted with NIB prior to use.

16. To ensure the blocking and labeling solutions keep in contact with the sample, either a plastic coverslip can be placed over the sample and solution, or a temporary well can be made by using a PAP pen to draw a well around the sample area. For in situ hybridization, the plastic coverslip should be made from a heat-resistant plastic such as an autoclave bag so that it is resistant to the high denaturation temperatures used.

17. There are many mounting solutions available, but it is important for optimal image collection to match refractive indexes as closely as possible within your imaging setup. Ideally the refractive index of the immersion medium for the lens (in this case oil), the glass coverslip, and the sample mounting medium should be the same. The mounting medium should also have good anti-fade properties and be able to limit the amount of fluorescence quenching through photobleaching. We have found that a solution of 97 % TDE in PBS pH 7.0 [6] is very good for cy3 and cy5 and other fluorochromes at these longer wavelengths. However the fluorescence of Alexa 488 is less stable and Vectashield is better for this fluorochrome.

18. Most objectives designed for use in high-resolution biological imaging are calculated for a coverslip thickness of 170 nm (No 1.5). For the highest quality imaging, we recommend using high-precision coverslips such as Carl Zeiss high-performance coverslips, as these have a much smaller deviation from the nominal 170 nm than standard coverslips.

19. An extra fixation step is sometimes included if the antigen/antibody complex is suspected to be unstable and could be disturbed by the conditions of the in situ hybridization. Over-fixation, however, can lead to penetration problems of probes and antibodies, so careful monitoring of this step is important.

## Acknowledgments

This work was supported by grant BB/J004588/1 from BBSRC and the John Innes Foundation. We thank Drs Azahara Martín and Stefanie Rosa for helpful comments on the manuscript.

## References

1. Hawes C, Satiat-Jeunemaitre B (eds) (2001) Plant cell biology: a practical approach. Oxford University Press, Oxford, p 338

2. Schwarzacher T, Heslop-Harrison P (2000) Practical in situ hybridization. Bios Scientific Publishers, Oxford, p 203

3. Prieto P, Moore G, Shaw P (2007) Fluorescence in situ hybridization on vibratome sections of plant tissues. Nat Protoc 2:1831–1838

4. Rawlins DJ, Highett MI, Shaw PJ (1991) Localization of telomeres in plant interphase nuclei by in situ hybridization and 3D confocal microscopy. Chromosoma 100:424–431

5. Collier S, Pendle A, Boudonck K, van Rij T, Dolan L, Shaw P (2006) A distant coilin homologue is required for the formation of cajal bodies in arabidopsis. Mol Biol Cell 17:2942–2951

6. Staudt T, Lang MC, Medda R, Engelhardt J, Hell SW (2007) 2, 2′-thiodiethanol: a new water soluble mounting medium for high resolution optical microscopy. Microsc Res Tech 70:1–9

# Chapter 7

# Manipulation of Homologous and Homoeologous Chromosome Recombination in Wheat

## Adam J. Lukaszewski

## Abstract

Given the sizes of the three genomes in wheat (A, B, and D) and a limited number of chiasmata formed in meiosis, recombination by crossing-over is a very rare event. It is also restricted to very similar homologues; the pairing homoeologous (*Ph*) system of wheat prevents differentiated chromosomes from pairing and crossing-over. This chapter presents an overview and describes several systems by which the frequency or density of crossing-over can be increased, both in homologues and homoeologues. It also presents the standard system of E.R. Sears for engineering alien chromosome transfers into wheat.

**Key words** *Triticum aestivum*, Crossing-over, The *Ph* system, Recombination stringency, Structural chromosome variants

## 1 General Comments

As in most eukaryotes, the levels of crossing-over in wheat are low, whether on the *per* genome, *per* chromosome, or *per* DNA base-pair basis. For proper chromosome segregation in anaphase I of meiosis, a single crossover event producing a chiasma involving two homologues is fully sufficient. As proper reduction of the chromosome number is the primary goal of meiosis, most organisms have developed mechanisms which limit the numbers of crossovers per chromosome, to assure that all chromosome pairs develop at least that critical one. This mechanism is the positive chiasma interference, which limits the probability of additional crossovers in the vicinity of the already established ones. In wheat, the physical distance of the positive chiasma interference is quite substantial, stretching to about one half of an average chromosome arm length [1]. As a consequence, physically long arms may develop second, additional crossovers/chiasmata; short arms are usually limited to just one. Given the total DNA contents of an average chromosome arm, this translates into single crossovers per

Shahryar F. Kianian and Penny M.A. Kianian (eds.), *Plant Cytogenetics: Methods and Protocols*,
Methods in Molecular Biology, vol. 1429, DOI 10.1007/978-1-4939-3622-9_7,
© Springer Science+Business Media New York 2016

many megabases of DNA. Hexaploid wheat, with its three genomes of seven chromosomes each (A, B, and D) and ca. 5000–6000 Mb of DNA per genome, forms ca. 40–50 chiasmata (crossovers) in an average meiocyte or around one every 300 Mb of DNA. For many experimental purposes, these crossover frequencies are too low, especially when the goal is to saturate specific regions with high numbers of exchanges, to break tight linkages or to generate intra-locus recombinants. In such cases, attempts must be made to increase crossover rates in the designated regions. This can be accomplished by making use of specific chromosome constructs. The author is not aware of any successful attempts to increase the crossover rate on the genome-wide basis. This would require experimental manipulation of the positive chiasma interference, quite a difficult proposition when the actual mechanism is unclear.

## 2  Changing Crossover Density in Designated Homologous Segments

Wheat, similar to many other eukaryotes, depends on the lepto-tene bouquet for homologue alignment [2]. In the bouquet stage, telomeres congregate in a tight configuration; this step immediately precedes the initiation of synapsis. Some other mechanisms for homologue alignment must also be present and operate at the same time, but its nature is not clear, and it appears to account for a very small proportion of successful homologue alignments [3]. Perhaps because the initiation of synapsis is telomeric, and so the distal chromosome segments are the first ones to be fully synapsed, most crossovers are located distally, in the vicinity of the telomere. The average frequency of crossovers drops quickly with distance toward the centromere. In chromosome arms with sufficient length, a spike of the crossover rate may appear in the middle of the arm [1]. There is little, if any, crossing-over in the proximal halves of the arms. For a while, it was believed that absence of crossovers reflected only the pattern of synapsis and timing (late synapsis in the vicinity of the centromere would preclude crossing-over). However, it turned out that the proximal regions of chromosome arms are physically incapable of forming crossovers, even when placed in the vicinity of the telomere and, hence, still the first to synapse [3, 4]. It is not clear what may be responsible for licensing chromosome segments for crossing-over.

With the mechanism of licensing unknown, it does not appear likely that accessing nonrecombining regions of chromosome arms will be possible anytime soon. This still leaves the distal regions, and these are the regions harboring most genes. Here, the manipulation is relatively simple: crossovers can be restricted to some specific region of the arm, and this increases their density. The critical segment is placed as close to the telomere as possible, by, for instance, making use of sets of deficiencies (chromosomes missing terminal segments). There are plenty of those in wheat and some in other species; and new ones can easily be made by making use of the so-called gametocidal chromosomes [5]. Since pairing initiation for a deficient chromosome arm is still telomeric, the segment close to the telomere will have increased crossover rate over that in its standard, intercalary position (Fig. 1). This system has been tested in two experiments in wheat, with chromosome arm 5BL deficient for ca. terminal 50% of the arm [6] and 1BL deficient for terminal 23% of the arm [7]. In both cases, the crossover rates of the new, terminal segments were greatly increased over their standard rates in normal intercalary positions. It is important to note that heterozygosity for deficiency breakpoints misaligns the two arms at the telomere bouquet and may drastically reduce chances for pairing and recombination. This dictates a very demanding step in stock preparation: allelic variation is required to detect recombination, so two chromosomes from different sources are required. Identical breakpoints cannot be produced by random breakage, so a breakpoint from one chromosome must be recombined onto the other chromosome but with the exchange point as close to the telomere of the deficiency chromosome as possible, to preserve allelic variation along the rest of the arm. Depending on the position of the breakpoint, this can be a challenging exercise.

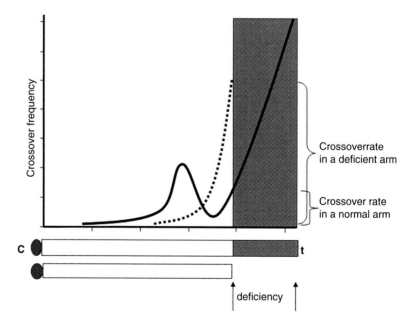

**Fig. 1** In a normal chromosome arm in wheat, crossing-over is concentrated in the terminal segment and in sufficiently long arms may form a second peak in the middle of the arm (*solid line*). An interstitial segment may recombine, but with a low frequency. If the chromosome arm is truncated by a deficiency, an intercalary segment is now located next to the telomere, and its crossover frequency increases dramatically. *C* is centromere, *t* is telomere

In a normal chromosome arm, telomeric initiation of pairing produces distal concentration of crossovers, but they are still distributed over a fairly large segment of the arm, perhaps as much as the distal one half. If the length of this segment is restricted on the proximal side, by a translocation or inversion, or some other obstruction to normal synapsis and pairing, crossovers will concentrate exclusively in the only segment of homology available (Fig. 2). In essence, this is tricking a chromosome segment into a much higher crossover rate than its normal length and position dictate. This can be quite effective; the highest increase the author observed in a terminal segment restricted by a proximal translocation was 17.4-fold relative to the same segment in a standard chromosome. Four- to sixfold increases are common. The total number of crossovers in the arm may drop, but their concentration in the designated segment is increased; the effect is achieved by restricting the length of the chromosome arm over which crossovers can legitimately spread. The rate of increase appears directly proportional to the length of the segment permitted to crossover: the shorter it is, the higher the rate of increase.

Non-meiotic tests of crossing-over show that its success rate depends on the presence of sufficient stretches of perfect base-pair

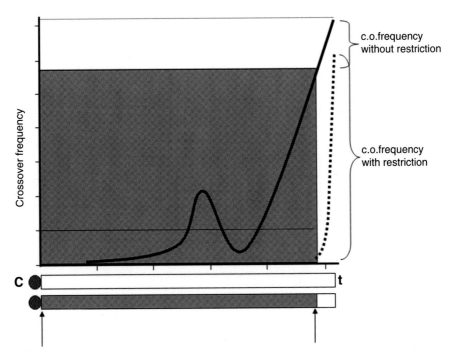

**Fig. 2** In a normal chromosome arm in wheat, crossing-over is concentrated in the terminal segment and in sufficiently long arms may form a second peak in the middle of the arm (*solid line*). If crossovers are restricted by a translocation (*between two arrows, shaded area*), the terminal segment will cross over with a frequency several times higher than in a normal chromosome arm (*broken line*). *C* centromere, *t* telomere

homology [8]. Single mismatches in the segment designated for crossing-over reduce the crossover rate by as much as four orders of magnitude [9]. At the same time, polymorphism is indispensable for detection of recombination. This implies that our picture of recombination may be distorted.

The *Ph* system of wheat appears capable of recognizing even subtle differences between chromosomes designated for pairing and recombination. In most cases, the *Ph* system is thought of as, and discussed in, the context of homoeologues and diploid-like meiosis in allopolyploids. However, it is quite sensitive and restrictive, and it also operates on homologues [10]. Therefore, certain stretches of DNA in homologous chromosomes may be more prone to (or capable of) recombination, while others may be less so. Given the low average number of crossovers per chromosome arm in wheat (say, 1.3–1.4 on the whole-arm basis), the effect of *Ph1* may be negligible because somewhere along the recombining portion of the arm, sufficient length of perfect homology can be found. In a restricted space, it may not. In an experiment with a small terminal segment of the T-9 wheat-rye recombinant chromosome [11] tested against six different homologues, disabling the *Ph1* system increased the average homologous crossover rate about 2.4-fold, reaching over a threefold increase in one instance (Lukaszewski, unpublished). This brought the level of recombination to that expected from the metaphase I pairing frequency of two identical segments (it is 50% [12], producing 25 cM of the genetic map length). Clearly, in the absence of the *Ph1* locus, DNA sequence differences between pairs of homologues are overlooked that otherwise would prevent crossover formation.

In summary, the frequency of homologous crossing-over in wheat can be increased significantly, by placing segments designated for recombination in terminal positions, by restricting recombination to short terminal segments, and by reducing the stringency of recombination. The only caveat of this approach is that the targeted segment must be capable of recombination in its standard position in a normal chromosome. In intervarietal hybrids of wheat, disabling of the *Ph1* locus may not increase the overall crossover rate but may make crossovers more evenly distributed along the arms by overlooking more polymorphic stretches of DNA. Whether nonrecombining segments of chromosomes can be forced to recombine is still an open question. Inversion of an arm, when nonrecombining regions normally close to the centromere are placed distally, did not induce these regions to form chiasma [13]. On the other hand, the distribution of crossover points in homoeologous pairs can at times be almost random on the centromere-telomere axis.

## 3  Recombination of Homoeologues

The *Ph* system in polyploid wheat restricts pairing (and crossing-over) to homologues. The distinction between homology and homeology is fluid. By "homoeologues," we understand genetically equivalent chromosomes belonging to different genomes. But the degree of difference may vary; it should be viewed as a continuum, from perfectly identical chromosomes such as those originating from sister chromatids to completely different chromosomes in different species. The *Ph1* system in wheat appears to control the stringency of crossing-over in such a manner that beyond some (very high) level of similarity, the probability of crossing-over drops off dramatically. The system at times prevents from pairing chromosomes in intervarietal hybrids even though by standard definitions these ought to be considered homologues [10].

Because of the *Ph* system, tetraploid and hexaploid wheats behave in meiosis as diploids, that is, they form only bivalents in metaphase I. When the *Ph* system is disabled, homoeologues are capable of pairing and multivalents result. Two major loci of the *Ph* system are known, *Ph1* on the long arm of chromosome 5B and the *Ph2* locus on the short arm of chromosome 3D, plus several minor loci on various other chromosomes (for review, *see* ref. 14). *Ph1* has a much stronger effect; disabling both loci in some cases has no discernible effect [14]. In some others, it may slightly increase the levels of homoeologous recombination [15]. Because the effect of *Ph2* is small, in all practical attempts at homoeologous recombination, only the *Ph1* locus is manipulated. Most commonly, the so-called mutation, *ph1b*, is used. It was produced by irradiation [16, 17] and was eventually found to be a deletion of ca. 1.5 Mbp of DNA from an intercalary position on chromosome arm 5BL [18]. The *Ph1* system works not only with homoeologues of wheat (genomes A, B, and D) but just as well on homoeologues from related species. For this reason, it is manipulated to effect transfers of alien chromatin into wheat.

Since *Ph1* is dominant, in experiments with homoeologous recombination, homozygous *ph1b* mutation is combined with a monosomic substitution of an alien chromosome for the designated wheat recipient chromosome or, if the arm location of the targeted locus is known, with a centric translocation of the alien chromosome arm. Centric translocation may provide some advantages over double monosomics, such as regular disjunction of the two chromosomes. On the other hand, making use of double monosomy may increase recovery rates of recombined chromosomes, especially on the male side. The average gamete inclusion rate of monosomic chromosomes in wheat is 25 % [19], which means that 37.5 % of gametes produced by a double monosomic are euploid vs. 100 % euploid gametes produced as a consequence of pairing. If all aneuploid gametes are eliminated by gametic selection, the effective recombination rate scored among progenies should increase almost threefold.

Alien homoeologues introduced into wheat have a tendency to recombine not only with their designated recipient chromosome but also with any of the other two available homoeologues. Lukaszewski [11] called it "recombination fidelity." It can be an important factor as predictions of population sizes for any given level of precision in chromosome engineering must be revised upward, to compensate for recombination infidelity. In engineering of the rye chromosome arm, 1RS in wheat recombination was set up between 1RS and 1BS, but only ca. 70% of recovered recombinants were of the two arms; the remaining 30% were of 1RS with 1AS and 1DS [11]. In recombination of 2RS with 2BS, 76% of recovered recombinants were of the two designated arms; the rest were 2RS with 2AS and 2DS [20]. In a much more frequently recombining 2RL, 90% of recovered recombinants were of targeted arm (2BL) and only 10% of 2AL and 2DL. The sample appears too small to draw binding conclusions about the relationship between homoeologue affinity and recombination fidelity.

The crossover rates of homoeologues in the absence of *Ph1* appear to be arm specific. They can vary greatly and appear to depend on two factors: the general affinity of the chromosome arms destined for recombination and their structural similarity. In wheat-rye recombination, metaphase I pairing frequencies of individual chromosome arms can range from zero to as high as 12% [21]. Arm specificity of crossover rates is best illustrated by chromosomes 1R and 2R. Chromosome arm 1RS, in a very large experiment set up to recombine it with 1BS, showed recombination rate of ca. 0.4%; when set up with 1DS, its recombination frequency was somewhat lower [11, 22]. The long arm, 1RL, recombined with its 1AL, 1BL, and 1DL homoeologues at 7.3% [23]. In chromosome 2R, the short arm recombined with its wheat homoeologous arms 2AS, 2BS, and 2DS with a combined frequency of. 0.3% (one recombination event per 322 progeny chromosomes), while the long arm recombined with ca. 16.3%. The low crossover rate of 2RS is understandable; relative to wheat, this arm is translocated, carrying in the terminal position short segments homoeologous to wheat group 6 and 7 chromosomes [24]. While its alignment with wheat 2AS, 2BS, and 2DS may still be normal (we do not know that), there is no homeology available in the vicinity of the telomere to support crossing-over. However, 1RS is syntenic with wheat group-1 short arms so its low crossover rate is more difficult to explain. Unless targeted chromosomes are well known and characterized, and most often they are not, it is impossible to make predictions about their behavior and, therefore, the success rate in a recombination attempt. If the transfer is to be done with high precision, one must be prepared to screen either truly large samples of progenies to make sure that all break-points flanking the desired locus are available after the first round of screening or proceed by consecutive steps of refining the transfers, as shown by the 22-year-long (as judged by the publication record) Australian effort [25–27]. For comparison, the direct approach of Lukaszewski [11] required about 8 years but far larger populations. On the other end of the spectrum, chromosomes of more closely related species, such as 5A*m* from *T. monococcum*, may recombine with their wheat homoeologues with a perfectly normal, homologous frequency, when *Ph1* is disabled [28, 29].

Chromosome arm 2RS illustrates how structural differences affect crossover frequency and distribution. Homologous recombination in wheat is distal, with longer arms showing regular second crossovers in intercalary positions, producing the average (in wheat) of ca. 1.3–1.4 crossovers per chromosome arm. Homoeologous recombination in the absence of *Ph1* tends to be limited to single events (but double crossovers have been noted), and the frequency depends on the level of chromosome affinity. Because the terminal segment in 2RS is homoeologous to terminal regions of the group-6 and 7 short arms [24], its alignment with 2S arms of wheat is disturbed, thus affecting the crossover rate. At the same time, 2RS does not appear to recombine with 6S and 7S arms of wheat; at least, no such recombinants were recovered among 8193 progenies [20]. For chromosome arms with more serious rearrangements, no useful recombinants of 4RL from *Secale montanum* in wheat were recovered among 3653 progenies [4], and no recombinants were recovered for rye chromosome 6RL from a population of ca. 3800 [30]. Surprisingly, the translocation breakpoints recovered for 2RS were distributed much more randomly on the centromere-telomere axis than is normally the case in homologous [1] and homoeologous recombination [23]. Perhaps misalignment of arms alters the pattern of crossover distribution, whether the *Ph1* locus defines which chromosome segments are capable of recombination and which are yet to be tested.

The *Ph1* system in wheat can be disabled in several ways. The *ph1b* mutation is available in cv. Chinese Spring where it can be identified by DNA markers [18] or meiotic pairing patterns; an alternative is a crossing scheme, which permits tracking of specific chromosomes, such as using double monosomics with 5B. In the first cross between an alien addition of the donor chromosome as male and a double monosomic, the alien chromosome is placed as a single chromosome substitution for its target wheat homoeologue, and 5B (with *Ph1*) is monosomic. This plant is backcrossed as female to the 5B*ph1b* line, and triple monosomic progenies are selected: monosomic alien donor, monosomic wheat recipient, and monosomic 5B*ph1b*. The same effect can be accomplished by making use of DNA markers permitting identification of the recipient and donor homoeologues and the *ph1b* mutation. The resulting plants will undergo homoeologous recombination, including that of the donor and recipient chromosomes. The author simplified this procedure some more: the 5B*ph1b* from Chinese Spring was transferred to cv. Pavon 76 where 5B*ph1b* can be unambiguously identified by its banding pattern. In most cases, C-banding also permits identification of the donor and recipient homologues, making the analysis a one-step procedure, probably no more complex and time-consuming than DNA extraction for marker analysis. However, the cytogenetic approach requires some minimum skills in cytology.

The *Ph1* locus can also be eliminated by 5B nullisomy: monosomic 5B pollinated with an alien donor species produces two types of $F_1$ hybrids: those with 5B present and no homoeologous pairing and nulli-5B with high homoeologous pairing. The author repeatedly recovered 5B nullisomics in Chinese Spring, and in their early generations, these are almost normally fertile and can be used in backcrosses both as male and as female. Unfortunately, high pairing in interspecific $F_1$ hybrids, as can be expected in the absence of *Ph1*, makes progeny recovery a very difficult task, and unless amphiploids are produced, it may be practically impossible to generate large-enough backcross progeny populations for a sensible chance of recovery of useful recombinants.

Perhaps the simplest approach to homoeologous recombination is suppression of the *Ph1* system by chromosomes from *Aegilops speltoides* [31]. The suppressors are dominant and, therefore, do not require as much stock preparation as the *ph1b* mutation, but there are too few published reports to assess the efficiency and parameters of this approach relative to the well-tested *Ph1* system [32].

## 4  The Two-Step Approach to Engineering Alien Chromosome Segments into Wheat

The system of chromosome engineering by homoeologous recombination was created by E.R. Sears [33] and is used to this day (Fig. 3). It consists of two steps: in the first step, primary recombinants are isolated from among progenies of plants with disabled *Ph1*. Since most are single crossover events, these primary recombinants appear in two configurations: with proximal wheat and distal alien segments and vice versa. Multiple crossovers per arm are rare and cannot be counted on to simplify the procedure. Recovered primary recombinants are screened for the presence of the desired locus and the positions of the translocation breakpoints. From each configuration, one is selected, with the desired locus present and the closest available breakpoint, combined in a single plant with *Ph1* present, and allowed to recombine. *Ph1* permits only homologous recombination, and the only segment of homology in the two

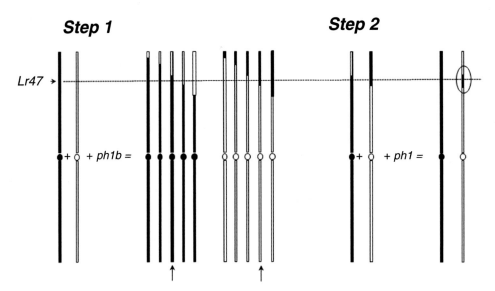

**Fig. 3** The two-step procedure of E.R. Sears for chromosome engineering, illustrated using a transfer of the *Lr47* locus from *A. speltoides* chromosome 7S (*solid*) to chromosome 7A of wheat (*shaded*). In the first step, 7S and 7A are combined with the *ph1b* mutation and recombined chromosomes are recovered. These are in two configurations: 7S with terminal segments of 7A and 7A with terminal segments of 7S. All chromosomes are screened for the presence of *Lr47* and the location of the translocation breakpoints. From each group, one is selected with the breakpoint closest to *Lr47* (*arrowed*) and in step 2 allowed to recombine in the presence of *Ph1*, producing a revertant alien (7S) chromosome and a wheat chromosome (7A) with intercalary introgressions (*circled*) from *A. speltoides*

primary recombinants is between the two translocation breakpoints. Any crossover in that segment produces two chromosomes: a wheat chromosome with an intercalary alien introgression and a normal alien chromosome (or arm) which is of little use (Fig. 4).

The ultimate size of the introgressed segment depends only on the positions of the two breakpoints selected for the second step of the procedure. Therefore, the precision of the entire exercise is determined by the number of primary recombinants recovered: the more that are isolated, the better the chance of finding two flanking the target locus in immediate vicinity. The number of primary recombinants in turn depends on the recombination frequency of the donor and target chromosome arms and the size of the screened population. If the recombination rate of the two arms is taken as a constant, the size of the screened population determines the precision of a transfer: The calculation is simple (metaphase I pairing rates for each rye chromosome in wheat are known [21]): with pairing rate of 2 % (recombination rate of 1 %) and with no more than 1 cM of alien chromatin to remain on either side of the targeted locus, ca. 30,000 progenies have to be screened for a 95 % probability of success [the formula is $n = \mathrm{Ln}(1-p)/\mathrm{Ln}(1-x)$ where $n$ is the population size, $p$ the desired probability of success, and $x$ the event frequency]. Unfortunately, there is nothing simple

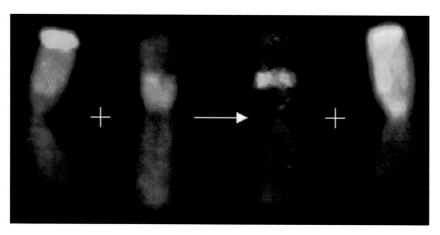

**Fig. 4** Production of an intercalary insert from rye into wheat chromosome 2B: two primary recombinants 2RS and 2BS in reciprocal configurations (*two on the left*) are allowed to recombine in the presence of *Ph1*. A crossover event in the segment of homology of the overlapping primary recombinants produces a wheat chromosome with an intercalary rye insert (*third from the left*) and a normal rye chromosome arm (*first on the right*). Rye chromatin labeled *green*; wheat chromatin in *red*

and quick about chromosome engineering. However, the process can perhaps be simplified somewhat by mass selection among progenies to identify plants with the desired locus, and absent are some other characteristics known to be associated with the presence of the donor alien chromosome.

Being so rare, primary chromosome recombinants are necessarily precious. This creates an interesting dilemma: should primary recombinants be generated for possible future uses, and should those already created become collection items to avoid the labor involved in their selection? The answer depends on the consideration of the effort needed to transfer new desirable loci as they are being identified (it is difficult to imagine selecting recombinants with resistance to yet unknown races of a pathogen), onto the existing primary recombinants. If precision is required, and small introgressions are desirable specifically because they are small, then the crossover rates in very narrow intervals (such as 1 cM flanking the targeted locus) will be low. In fact, they are much lower than the genetic lengths of the existing segments imply. Differences in chromosome configuration (such as between a donor alien chromosome and a recombinant with a terminal wheat segment) interfere with synapsis and seriously reduce the effective crossover rate. Depending on the chromosome in question, this reduction may be three- to fourfold [11, 12]. To illustrate, if no more than 1 cM worth of alien chromatin flanking the locus of interest is desired and only one specific recombinant is needed (of the two possible), the expected recovery rate based on genetic distance is 0.5 % or 1 in 200. For a 95 % probability of recovery of the desired construct, ca. 600 progenies would have to be screened. If these frequencies are reduced three- to fourfold by the difference in chromosome configuration, the actual population will be in the range of at least 1800–2400, not a trivial job. Depending on the pairing/recombination frequency of the donor alien chromosome with one of its wheat homoeologues, this may be less (but also much more) than the population needed to recover the desired primary recombinant from direct crossover between the donor and recipient in *ph1b* condition. And this is only one of two chromosomes needed to produce an intercalary introgression. So, the existing populations of recombinant chromosomes can be used in the future to create desired small-segment introgressions, but the effort will still require screening large populations.

## References

1. Lukaszewski AJ, Curtis CA (1993) Physical distribution of recombination in B-genome chromosomes of tetraploid wheat. Theor Appl Genet 86:121–127

2. Corredor E, Lukaszewski AJ, Pachón P et al (2007) Terminal regions of wheat chromosomes select their pairing partners in meiosis. Genetics 177:609–706

3. Lukaszewski AJ (2008) Unexpected behavior of an inverted rye chromosome arm in wheat. Chromosoma 117:569–578

4. Lukaszewski AJ, Porter DR, Baker CA et al (2001) Attempts to transfer Russian wheat aphid resistance from a rye chromosome in Russian triticales to wheat. Crop Sci 41:1743–1749

5. Endo TR, Gill BS (1996) The deletion stocks of common wheat. J Hered 87:295–307

6. Qi LL, Friebe B, Gill BS (2002) A strategy for enhancing recombination in proximal regions of chromosomes. Chromosome Res 10:645–654

7. Jones LE, Rybka K, Lukaszewski AJ (2002) The effect of a deficiency and a deletion on recombination in chromosome 1BL in wheat. Theor Appl Genet 104:1204–1208

8. Rubnitz J, Subramani S (1984) The minimum amount of homology required for homologous recombination in mammalian cells. Mol Cell Biol 4:2253–2258

9. Datta A, Hendrix M, Lipsitch M, Jinks-Robertson S (1997) Dual roles for DNA sequence identity and the mismatch repair system in the regulation of mitotic crossing-over in yeast. Proc Natl Acad Sci U S A 94:9757–9762

10. Dvorak J, McGuire P (1981) Nonstructural chromosome differentiation among wheat cultivars with special reference to differentiation of chromosomes in related species. Genetics 97:391–414

11. Lukaszewski AJ (2000) Manipulation of the 1RS.1BL translocation in wheat by induced homoeologous recombination. Crop Sci 40:216–225

12. Valenzuela NT, Perrera E, Naranjo T (2013) Identifying crossover-rich regions and their effect on meiotic homologous interactions by partitioning chromosome arms of wheat and rye. Chromosome Res 21:433–445

13. Lukaszewski AJ, Kopecky D, Linc G (2012) Inversions of chromosome arms 4AL and 2BS in wheat invert the patterns of chiasma distribution. Chromosoma 121:201–208

14. Sears ER (1976) Genetic control of chromosome pairing in wheat. Annu Rev Genet 10:31–51

15. Ceoloni C, Donini P (1993) Combining mutations for the two homoeologous pairing suppressor genes *Ph1* and *Ph2* in common wheat and in hybrids with alien Triticeae. Genome 36:377–386

16. Sears ER (1975) An induced homoeologous pairing mutant in wheat. Genetics 80:s74

17. Sears ER (1984) Mutations in wheat that raise the level of meiotic chromosome pairing. In: Gustafson JP (ed) Gene manipulation in plant improvement. 16th Stadler Genetics Symp., Columbia, MO. Plenum Press, New York. p 295–300

18. Gill KS, Gill BS, Endo TR, Mukai Y (1993) Fine mapping of *Ph1*, a chromosome pairing regulator gene in polyploid wheat. Genetics 134:1231–1236

19. Sears ER (1954) The aneuploids of common wheat. Mo Agr Exp Sta Res Bull 572:1–58

20. Lukaszewski AJ, Rybka K, Korzun V et al (2004) Genetic and physical mapping of homoeologous recombination points involving wheat chromosome 2B and rye chromosome 2R. Genome 47:36–45

21. Naranjo T, Fernandez-Rueda P (1996) Pairing and recombination between individual chromosomes of wheat and rye in hybrids carrying the *ph1b* mutation. Theor Appl Genet 93:242–248

22. Lukaszewski AJ (2006) Cytogenetically engineered rye chromosomes 1R to improve breadmaking quality of hexaploid triticale. Crop Sci 46:2183–2194

23. Lukaszewski AJ (1995) Physical distribution of translocation breakpoints in homoeologous recombinants induced by the absence of the *Ph1* gene in wheat and triticale. Theor Appl Genet 90:714–719

24. Devos KM, Atkinson MD, Chinoy CN et al (1993) Chromosomal rearrangements in the rye genome relative to wheat. Theor Appl Genet 85:673–680

25. Anugrahwati DR, Shepherd KW, Verlin DC et al (2008) Isolation of wheat-rye 1RS recombinants that break the linkage between the stem rust resistance gene *SrR* and secalins. Genome 51:341–349

26. Koebner RMD, Shepherd KW (1986) Controlled introgression to wheat of genes from rye chromosome 1RS by induction of allosyndesis. I. Isolation of recombinants. Theor Appl Genet 73:197–208

27. Rogovsky PM, Gudet FLY, Langride P et al (1991) Isolation and characterization of wheat-rye recombinants involving chromosome arm 1DS of wheat. Theor Appl Genet 82:537–544

28. Luo MC, Dubcovsky J, Dvorak J (1996) Recognition of homoeology by the wheat *Ph1* locus. Genetics 144:1195–1203

29. Luo MC, Yang ZL, Kota RS, Dvorak J (2000) Recombination of chromosomes 3A$^m$ and 5A$^m$ of *Triticum monococcum* with homoeologous chromosomes 3A and 5A of wheat: distribution of recombination across chromosomes. Genetics 154:1301–1308

30. Dundas IS, Frappell DE, Crack DM, Fisher JM (2001) Deletion mapping of a nematode resistance gene on rye chromosome 6R in wheat. Crop Sci 41:1771–1778

31. Dvorak J (1972) Genetic variability in *Aegilops speltoides* affecting homoeologous pairing in wheat. Can J Genet Cytol 14:317–380

32. Chen PD, Tsujimoto H, Gill BS (1994) Transfer of *Ph1* genes promoting homoeologous pairing from *Triticum speltoides* to common wheat. Theor Appl Genet 88:97–101

33. Sears ER (1981) Transfer of alien genetic material to wheat. In: Evans L, Peacock WJ (eds) Wheat science – today and tomorrow. Cambridge University Press, Cambridge, pp 75–89

# Chapter 8

# Dissecting Plant Chromosomes by the Use of Ionizing Radiation

Penny M.A. Kianian, Katie L. Liberatore, Marisa E. Miller, Justin B. Hegstad, and Shahryar F. Kianian

## Abstract

Radiation treatment of genomes is used to generate chromosome breaks for numerous applications. This protocol describes the preparation of seeds and the determination of the optimal level of irradiation dosage for the creation of a radiation hybrid (RH) population. These RH lines can be used to generate high-resolution physical maps for the assembly of sequenced genomes as well as the fine mapping of genes. This procedure can also be used for mutation breeding and forward/reverse genetics.

**Key words** Radiation hybrid, Gamma rays, Tempering seed, Mutation, Mapping, Gene cloning

## 1 Introduction

Since the pioneering work of geneticist Hermann Muller, who utilized X-rays in the early 1900s to mutagenize *Drosophila melanogaster* [1], ionizing radiation has been used as a powerful source for inducing mutations in animal and plant experimental systems [2, 3]. Different dosages and sources [e.g., ultraviolet (UV), X-rays, and γ-rays] can be utilized to induce mutations of varying degrees of severity, including point mutations, rearrangements, and deletions (Fig. 1). Because DNA repair pathways have different levels of fidelity [3, 4], mutations generated via radiation can persist in the genome and may be harnessed for a variety of experimental applications. Ionizing radiation (IR) such as X-ray and γ-rays lead to the formation of high levels of reactive oxygen species (ROS) due to the decomposition of water into hydroxyl radicals and hydrogen peroxide, which can damage all cellular constituents including DNA [3, 5, 6]. X-rays and γ-rays are orders of magnitude higher in energy than UV radiation and typically cause more single- and double-strand breaks (SSB and DSB, respectively) than UV-B [2] leading to small deletions, large

Shahryar F. Kianian and Penny M.A. Kianian (eds.), *Plant Cytogenetics: Methods and Protocols*,
Methods in Molecular Biology, vol. 1429, DOI 10.1007/978-1-4939-3622-9_8,
© Springer Science+Business Media New York 2016

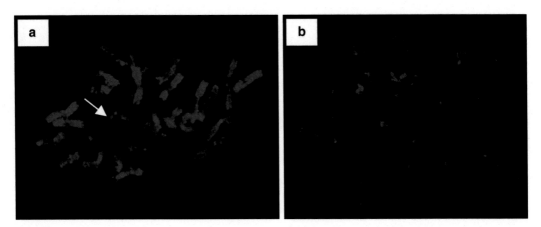

**Fig. 1** Cytological analysis of chromosome breaks in *Triticum turgidum* ssp. *durum* chromosome substitution lines after irradiation. The D-genome was probed with *T. tauschii* genomic DNA labeled with Rhodamine (*red*), while the A and B genomes were blocked with *T. turgidum* DNA and revealed using DAPI counterstain (*blue*). (**a**) Translocation between an alloplasmic durum line (lo) with 1AL.1D*scsae* chromosome [7] and a new translocation indicated with an *arrow*; (**b**) substitution line Langdon 1D(1A) [8] with interstitial deletions detected by molecular markers

chromosome deletions, translocations, and inversions. Once a DSB has formed, it can be repaired through one of two pathways: homologous recombination (HR) or non-homologous end joining (NHEJ). HR is a repair process that uses an undamaged copy of the broken region as a template to repair the break. The intact copy is usually supplied by a sister chromatid but can also be located on a homologous chromosome. Therefore, HR is generally an error-free process. In NHEJ, the ends of the single-strand chains are joined without requiring sequence homology, and extensive deletions or insertions can occur [2–4]. This error-prone repair process can be exploited to introduce novel genetic polymorphism and is central to the use of radiation to further understand and manipulate plant and animal genomes [7, 8].

Mutants with novel phenotypes resulting from radiation treatment have provided a rich source of material for further investigation via classical forward genetic screens and gene cloning approaches. Agronomically valuable mutants have also been developed in plant mutant breeding programs. Additionally, IR sources, most commonly γ-ray and X-ray, have been used in both animal and plant systems to generate "radiation hybrid" (RH) populations and have proven useful for a number of applications including improvements to physical chromosome maps, gene cloning, and most recently genome assembly [3].

Radiation hybrid techniques were first used in the 1970s by scientists seeking to generate hybrid chromosomes between human and rodent cells in mammalian tissue culture systems [9]. The technique has since been utilized for a number of animal and plant species. Briefly, in mammalian cell culture systems, one cell type is irradiated, inducing DSBs across the genome, and is then "rescued" by fusion with a second cell type in which NHEJ repair of DSBs results in hybrid chromosomes. Similarly in plants, one species is irradiated and crossed to a second "rescuer" species to generate a hybrid [10]. This technique is advantageous, because in contrast to native recombination events, radiation generates a more uniform distribution of strand breaks across the chromosomes. Moreover, the extent of strand breakage, and therefore mapping resolution, can be controlled by radiation dosage. Thus, RH populations overcome a major limitation of recombination inherent to classical mapping populations. While not yet commonly utilized in plants, this method has the potential to be particularly useful for physical chromosome mapping and genome assembly in plants, overcoming the challenges of large genome sizes and uneven recombination frequencies characteristic of many important model and crop species.

The γ-ray irradiation protocol presented here was developed using several different Poaceae species. This protocol explains how to temper seed, calculate radiation dosage, and determine the optimal dosage for use in RH mapping of individual chromosomes or whole genomes.

## 2  Materials

1. Balance.

2. 60 % (v/v) glycerol solution. Prepare 60 % (v/v) glycerol solution by thoroughly mixing 600 ml of reagent grade glycerol and 400 ml ddH$_2$O.

3. Wide-mouth container(s) for 60 % glycerol solution (such as 6 oz. wide-mouth jars).

4. Tempering container (an airtight container such as a bell jar or large sealable plastic container with a rubber sealing gasket and a lock-down mechanism, e.g., an 8 quart consumer-grade storage container).

5. Shallow, open-mouth containers for seeds (such as large petri dishes).

6. Radiation source.

7. Seeds.

8. Container for irradiating of seeds (e.g., 50 ml plastic centrifuge tube, or repurposed wide-mouth plastic chemical container).

## 3    Methods

### 3.1    Seed Preparation

High-quality seeds (e.g., high germination rate and good embryo/ endosperm quality) are essential for the success and consistency of results (*see* **Note 1**). The seed stock starting material will depend on your objectives (*see* **Note 2**).

### 3.2    Seed Tempering

Seeds are tempered in a controlled, hydrated environment to establish uniformly high moisture content within the seed. This provides adequate water within the plant cell for free radical production upon hydrolysis with ionizing radiation. Free radicals cause the chromosome breakage events.

#### 3.2.1    Tempering Process

1. Add glycerol solution to wide-mouth container(s) and write the total mass on the vessel (e.g., 450 g). If multiple containers are used to facilitate timely equilibration, add glycerol to a common, uniform mass. In the process of tempering, water will be lost from the glycerol solution via evaporation; consequently, water will need to be added to maintain a constant 60 % (v/v) concentration. Water is added to bring the total vessel mass back to the mass indicated on the vessel.

2. Place the wide-mouth vessel(s) with glycerol solution into the tempering container (Fig. 2); this creates a larger surface area from which water can evaporate out of the glycerol solution and into the tempering seeds. It is recommended to have a 10 % minimum ratio of glycerol solution volume to tempering container volume.

**Fig. 2** Example of a seed tempering setup using a bell jar. An open, wide-mouth glass jar containing 60 % (v/v) glycerol is placed at the *bottom* of the bell jar. Seeds are placed in an open glass petri dish above the glycerol solution and the bell jar is sealed. If multiple seed stocks are tempered at one time, a larger storage container and multiple glycerol containers may be necessary (see text)

3. Record (1) the total mass of each glycerol container, (2) the total mass of the seed to be irradiated, and (3) the initial tempering date.

4. Place the open (cap removed) glycerol container(s) into the tempering container.

5. Spread the seeds to be tempered evenly in a layer (1–3 seeds deep) in a shallow, open-mouth container such as a petri dish and record the mass of the container and seeds. 100 g of seed per tempering container is a practical upper limit. Place the seeds in the tempering container along side or above the open glycerol container.

6. Seal the tempering container.

7. After 4 days, check the mass of the glycerol vessel(s) and seed, and record those values with the date. Net loss of mass from the glycerol vessels and net gain to the tempering seeds should be evident. There will also be some unaccounted loss of water to the environment, presumably through the seal, or to the air in the tempering container. Add ddH$_2$O to the glycerol vessel(s) to bring them back to their indicated masses.

8. Repeat this process every day or two until the seed moisture content equilibrates. The seed will continue to temper and increase in mass, albeit at a diminishing rate. Generally, seed moisture content equilibrates after 7–10 days at about 13 %.

9. Transfer tempered seed to an airtight container (e.g., 50 ml plastic centrifuge tube or plastic bottle) that fits into the irradiator.

10. Irradiate (*see* **Note 3**).

**3.3 Sample Irradiation**

Seed irradiation (*see* **Note 4**) via γ-rays requires access to a very high-energy gamma irradiator. Typically, such irradiators use cesium-137 or cobalt-60 as a radiation source. Irradiators are generally located at large research universities and federal research facilities.

Irradiation dosage is a function of the source's specific activity, the time of exposure, and the distance between the source and the material. Generally, the physical design of the irradiators creates a fixed exposure distance, and thus the only variables in calculating dosage are the specific activity of the source and the time of exposure. Activity is the rate of dosage per unit time. Current activity for any fixed-distance, fixed-rate source is determined from the equation $A = A_0 e^{(-0.69315 \times t)} T^{1/2}$, where $A$ = current source activity; $A_0$ = original source activity; $t$ = time from $A_0$, in years; and $T_{1/2}$ = isotope half-life, in years. For example, consider a 3200 Curie cesium-137 (Cs$^{137}$) source calibrated at 1310 gray (Gy)/h on 14 July 1970. The half-life for Cs$^{137}$ is 30.07 years. The time elapsed from $A_0$ (14 July 1970) on 1 Jan 2015 would be 44.5 years, and so the corresponding activity would be 469.67 Gy/h. Therefore, a 100 Gy dosage on 1 Jan 2015 would be conferred by the source onto a sample in 12 h and 45 min.

**3.4 Optimal Irradiation Dosage**

Optimal dosage is calculated using seedling survival rate (*see* **Note 5**). To find the optimal radiation dosage, a range of dosages is tested. The goal is to create mutations without significantly decreasing plant survival. Dosages typically tested range from 0 to 500 Gy (*see* **Note 6**).

1. Divide tempered seeds into groups of 10–50 seeds for exposure to a range of radiation dosages.

2. Expose seed for the calculated time based on desired dosage. Using increments of 50 Gy is recommended for generating a dosage curve.

3. Irradiated seeds can be pre-germinated and then transplanted, or planted directly into soil. Note the treatment dosage applied to each seed.

4. For each dosage group, count the number of surviving seedlings about 2 weeks after date of planting or at the time when seedling growth is no longer dependent on the seed endosperm for nutrients (Fig. 3).

5. The number of surviving seedlings compared to the number of seedlings from the untreated control is the survival rate at each dosage, represented as a percentage.

**Fig. 3** Comparison of seedling survival and growth at several irradiation dosages. Seedlings germinated from seeds with the lowest irradiation dosage (200 Gy, *far left*) to higher irradiation dosages in 50 Gy increments (up to 350 Gy, *far right*) are shown. Seedling survival and growth rate decreases with increasing irradiation, as expected

**Table 1**
**Example of the data collected to determine optimal irradiation dosage based on seedling survival**

| Treatment (Gy) | Number of seeds | Number of seedlings | Survival rate (%) |
| --- | --- | --- | --- |
| 0 | 50 | 49 | 100 |
| 50 | 50 | 50 | 100 |
| 100 | 50 | 48 | 98 |
| 150 | 50 | 47 | 96 |
| 200 | 50 | 49 | 100 |
| 250 | 50 | 45 | 91 |
| 300 | 50 | 33 | 67 |
| 350 | 50 | 27 | 55 |
| 400 | 50 | 30 | 61 |
| 450 | 50 | 25 | 51 |
| 500 | 50 | 10 | 20 |

6. The dosage chosen for future experiments is the level when an obvious decrease in survival rate is observed. A dosage of 250 Gy is the optimum level in Table 1. A survival rate as low as 40–60 % may be optimum depending on the researcher's goals and the plant species.

*3.5   Seed Survival Applied to Population Development*

1. Using the information gained from Subheading 3.4, the experimental tempered seeds are irradiated. The seed number and dosage is dependent on the purpose of the experiment (*see* **Notes 7** and **8**).

2. Seeds are irradiated at the desired dosage(s) (*see* **Notes 9** and **10**), and plants are grown from the irradiated seed.

3. The mutant plants are hemizygous for their mutations. These plants can be selfed to generate homozygous lines for use in forward/reverse genetic screens or in mutational breeding. Alternatively, mutated heterozygous plants can be crossed to a non-irradiated parent to create a radiation hybrid (RH) mapping panel.

# 4   Notes

1. We recommend a seed germination test on seeds of unknown quality before irradiation, and regeneration of seed if necessary. Low-quality seeds make it difficult to determine if poor

germination and growth is due to radiation or is an effect of poor seed quality.

2. For mapping genes, chromosomes, or genomes, cytological stocks such as addition, deletion, substitution, and/or translocation lines allow for analysis of one chromosome at a time. This can simplify polyploid genome analysis in regions of high homology between and within a genome. Whole genome radiation of aneuploid stock can be done for polyploid species if molecular markers are specific for the polyploid genomes (*see* ref. [11, 12]).

3. Tempered seed should be irradiated immediately, as increased moisture content can affect long-term seed viability. Irradiated seed should be planted immediately as well. If seed is not planted immediately, it should be dried to proper storage moisture content.

4. Efforts to irradiate pollen of grasses were attempted in order to decrease the time necessary to create irradiated material for genotyping. Radiation dosages of 10–30 Gy using 0.5 Gy intervals were applied with some successes in wheat, barley, rye, wild wheat relatives, and cotton. Generally, this approach does not result in highly viable plants making forward/reverse genetic screens difficult. However, this approach allows for early DNA sampling, and saturation of a genomic region with markers. Briefly, the method entails:

   (a) Prior to dehiscence, whole anthers (containing pollen) are collected into a plastic container and then irradiated.

   (b) The irradiated pollen is then applied to the stigma of a receptive female, which was emasculated 2 days prior, by shaking the anther or brushing pollen onto the stigma. The cross is isolated from contaminating pollen by covering the inflorescence using a glassine bag.

   (c) Three days after crossing, a 2,4 D solution (2,4-dichlorophenoxyacetic acid, 100 mg/l in water) is sprayed on the inflorescence to induce embryo development, then rebagged with a glassine bag.

   (d) After 3 days, spray daily with 25 ppm gibberellic acid solution (dissolve 100 mg gibberellic acid in 5–10 ml alcohol, then dilute in the appropriate amount of water to create a 25 ppm solution, store at 4 °C) until embryo collection.

   (e) At 10 days after crossing, a second treatment of 2,4 D solution is sprayed on the inflorescence, then rebagged with a glassine bag.

   (f) Embryos are collected from developing seeds 21 days after crossing. The endosperm can be used for DNA extraction and molecular marker analysis. Embryo culture can be utilized to regenerate whole plants.

5. Yonezawa and Yamagata [13] define optimum dosage as the level of radiation producing 10% chlorophyll mutants. Some researchers use the seedling "kill rate" instead of the seedling survival rate. The kill rate is calculated as the percentage of seedlings that do not survive. Others have used the midpoint where the kill curve (generated using the data in Table 1) is at its steepest decline.

6. Plant materials tested using this method include hexaploid and durum wheat, maize, sorghum, and numerous cytogenetic stocks (wheat substitution lines, wheat-barley chromosome additions, oat-maize chromosome additions). RH mapping has also been applied for barley and cotton (*see* ref. [14–18]).

7. For RH panel development for a single chromosome typically 100–300 seeds or individuals are used (*see* ref. [8, 18, 19]), whereas 600–1000 seeds are necessary for fine mapping of chromosome specific lines (*see* ref. [20, 21]). Whole genome mapping is recommended for species capable of interspecific hybridization such as wheat, maize, and cotton using genome-specific molecular markers distributed throughout the genome (*see* ref. [3]). Previous work with the D-genome of wheat used more than 1500 plants (*see* ref. [12]).

8. This protocol can also be used for mutational breeding and forward/reverse genetic purposes (*see* ref. [8, 20, 22, 23]). In these scenarios, the population size will be dependent on the genetically effective cell number (GECN) and the mutation rate of the irradiation dosage applied. The genetically effective cell number is the number of cells within the seed which will produce tissue responsible for meiosis/gamete production. GECNs range from two to six in plants, as summarized in Table 12.1 in Kumar et al. [3]. Species with different GECNs will vary with respect to the number of mutations that are passed via the gametes to the next generation. For instance, *Arabidopsis thaliana* has a GECN of 2, resulting in a recommended irradiated population size of 2000–3000 individual seeds to generate a single mutation (*see* ref. [24]). To specifically target a gene of interest in *Arabidopsis* with a 99% likelihood of mutation, it is estimated that a population of 85,000 lines is required (*see* ref. [25]).

9. For the development of RH maps, we recommend using at least two different dosage levels (*see* ref. [19, 20, 26–28]) for RH panel development. One panel should be generated using the optimal irradiation dosage (Subheading 3.4), and the second panel at a lower dosage (generating fewer breaks) to assist in assembling the RH map.

10. To increase the resolution of a RH map, the radiation dosage can be raised to generate more chromosome breaks (*see* ref. [10, 17]).

## Acknowledgments

Funding from the National Science Foundation, Plant Genome Research Program (NSF-PGRP) grant No. IOS-0822100 to SFK is gratefully acknowledged. Numerous students and postdoctoral scientists participate in the work on radiation treatments of wheat, barley, maize and sorghum at North Dakota State University to list here but are named among the many publications referenced in this article.

## References

1. Muller HJ (1927) Artificial transmutation of the gene. Science 66:84–87

2. Rastogi RP, Richa, Kumar A et al (2010) Molecular mechanisms of ultraviolet radiation-induced DNA damage and repair. J Nucleic Acids 2010:592980

3. Kumar A, Bassi FM, de Jimenez MKM et al (2014) Radiation hybrids: a valuable tool for genetic, genomic and functional analysis of plant genomes. In: Tuberosa R, Graner A, Frison E (eds) Genomics of plant genetic resources. Springer, Netherlands, Dordrecht, pp 285–318

4. Britt AB (1999) Molecular genetics of DNA repair in higher plants. Trends Plant Sci 4:20–25

5. Valko M, Rhodes CJ, Moncol J et al (2006) Free radicals, metals and antioxidants in oxidative stress-induced cancer. Chem Biol Interact 160:1–40

6. Wi SG, Chung BY, Kim JS et al (2007) Effects of gamma irradiation on morphological changes and biological responses in plants. Micron 38:553–564

7. Joppa LR, Williams ND (1988) Langdon durum disomic substitution lines and aneuploid analysis in tetraploid wheat. Genome 30:222–228

8. Hossain KG, Riera-Lizarazu O, Kalavacharla V et al (2004) Radiation hybrid mapping of the species cytoplasm-specific (scs^ae) gene in wheat. Genetics 168:415–423

9. Goss SJ, Harris H (1975) New method for mapping genes in human chromosomes. Nature 255:680–684

10. Riera-Lizarazu O, Vales MI, Ananiev EV et al (2000) Production and characterization of maize chromosome 9 radiation hybrids derived from an oat-maize addition line. Genetics 156:327–339

11. Riera-Lizarazu O, Leonard JM, Tiwari VK et al (2010) A method to produce radiation hybrids for the D-genome chromosomes of wheat (*Triticum aestivum* L.). Cytogenet Genome Res 129:234–240

12. Kumar A, Simons K, Iqbal MJ et al (2012) Physical mapping resources for large plant genomes: radiation hybrids for wheat D-genome progenitor *Aegilops tauschii*. BMC Genomics 13:597

13. Yonezawa K, Yamagata Y (1977) On the optimum mutation rate and optimum dose for practical mutation. Euphytica 26:413–426

14. Wardrop J, Snape J, Powell W et al (2002) Constructing plant radiation hybrid panels. Plant J 31:223–228

15. Wardrop J, Fuller J, Powell W et al (2004) Exploiting plant somatic radiation hybrids for physical mapping of expressed sequence tags. Theor Appl Genet 108:343–348

16. Gao W, Chen ZJ, Yu JZ et al (2004) Wide-cross whole-genome radiation hybrid mapping of cotton (*Gossypium hirsutum* L.). Genetics 167:1317–1329

17. Gao W, Chen ZJ, Yu JZ et al (2006) Wide-cross whole-genome radiation hybrid mapping of the cotton (*Gossypium barbadense* L.) genome. Mol Genet Genomics 275:105–113

18. Kalavacharla V, Hossain K, Gu Y et al (2006) High-resolution radiation hybrid map of wheat chromosome 1D. Genetics 173: 1089–1099

19. Mazaheri M, Kianian PMA, Kumar A, et al (2015) Radiation Hybrid Map of Barley Chromosome 3H. The Plant Genome doi:10.3835/plantgenome2015.02.0005

20. Michalak De Jimenez MK, Bassi FM, Ghavami F et al (2013) A radiation hybrid map of chromosome 1D reveals synteny conservation at a wheat speciation locus. Funct Integr Genomics 13:19–32

21. Bassi, FM, Ghavami F, Hayden, MJ et al (2016) Fast-forward genetics by radiation hybrids to saturate the locus regulating nuclear-cytoplas-

mic compatibility in Triticum. Plant Biotechnology Journal doi:10.1111/pbi.12532

22. Kynast RG, Okagaki RJ, Rines HW et al (2002) Maize individualized chromosome and derived radiation hybrid lines and their use in functional genomics. Funct Integr Genomics 2:60–69

23. Bassi FM, Kumar A, Zhang Q et al (2013) Radiation hybrid QTL mapping of *Tdes2* involved in the first meiotic division of wheat. Theor Appl Genet 126:1977–1990

24. Page DR, Grossniklaus U (2002) The art and design of genetic screens: *Arabidopsis thaliana*. Nat Rev Genet 3:124–136

25. Li X, Zhang Y (2002) Reverse genetics by fast neutron mutagenesis in higher plants. Funct Integr Genomics 2:254–258

26. Gyapay G, Schmitt K, Fizames C et al (1996) A radiation hybrid map of the human genome. Hum Mol Genet 5:339–346

27. Stewart EA, McKusick KB, Aggarwal A et al (1997) An STS-based radiation hybrid map of the human genome. Genome Res 7:422–433

28. Olivier M, Aggarwal A, Allen J et al (2001) A high-resolution radiation hybrid map of the human genome draft sequence. Science 291:1298–1302

# Chapter 9

# Optical Nano-mapping and Analysis of Plant Genomes

**Ming-Cheng Luo, Karin R. Deal, Armond Murray, Tingting Zhu, Alex R. Hastie, Will Stedman, Henry Sadowski, and Michael Saghbini**

## Abstract

Application of optical mapping based on BioNano Genomics Irys® technology (http://www.bionanogenomics.com/) is growing rapidly since its debut in November 2012. The technology can be used to facilitate genome sequence assembly and analysis of genome structural variations. We describe here the detailed protocol that we used to generate a whole genome BioNano map for *Aegilops tauschii*, the D genome progenitor of hexaploid wheat (*Triticum aestivum*). We are using the whole genome BioNano map to validate sequence assembly based on the next-generation sequencing, order sequence scaffolds, and ultimately build pseudomolecules for the genome.

**Key words** BioNano genome map, Optical mapping, Physical map

## 1 Introduction

The optical mapping technique was developed by Schwartz et al. [1] and has been used to construct ordered, genome-wide, high-resolution restriction maps from single, stained molecules of DNA. By visually mapping the location of restriction enzyme sites along a large DNA stretch of an organism, the spectrum of resulting fragments collectively serve as a unique "fingerprint" for that portion of the genome. With the recent development of the Irys platform along with the IrysChip V2 (http://www.bionanogenomics.com/), enabling separation of linear DNA molecules (Fig. 1) using fluidics and visualizing restriction enzyme sites using fluorescence labeling and microscopy, the technique has become widely accessible. The visualized linear molecules are then assembled into larger molecules by the computer system and software. This technology can be used by genome sequence projects interested in (1) detecting and correcting errors in previously assembled contigs and scaffolds and/or (2) ordering and orienting sequence scaffolds, estimating gap sizes between scaffolds, and creating super-scaffolds and building pseudomolecule(s) for genome assembly (Fig. 2).

Shahryar F. Kianian and Penny M.A. Kianian (eds.), *Plant Cytogenetics: Methods and Protocols*,
Methods in Molecular Biology, vol. 1429, DOI 10.1007/978-1-4939-3622-9_9,
© Springer Science+Business Media New York 2016

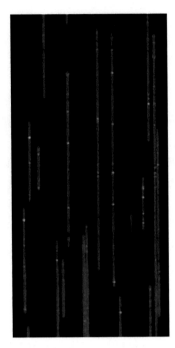

**Fig. 1** Linearized DNA molecules in nanochannels. The DNA molecule is stained with YOYO-1 (*blue*), and Nt.BspQI nicks are labeled with *green*

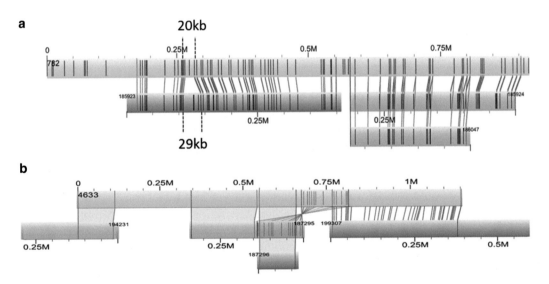

**Fig. 2** Alignment of sequence scaffolds (*blue bars*) generated from the next generation sequencing (NGS) on to BioNano contigs (*green bars*). (**a**) Extra sequences were assembled in one of the NGS scaffolds; ca. 9 kb sequences were inserted incorrectly during assembly. (**b**) An error in a NGS scaffold was detected. Two NGS contigs were mistakenly linked together with one of them in reversed order during the scaffolding process

In our case, a bacterial artificial chromosome (BAC)-based physical map was constructed for *Aegilops tauschii*, the D genome progenitor of bread wheat (*Triticum aestivum*) [2]. From this physical map, sequencing of the minimum-tiling-path (MTP) BAC clones (http://aegilops.wheat.ucdavis.edu/ATGSP/) will generate a reference sequence. The BioNano map for the genome was constructed to facilitate the genome assembly.

In the current protocol, we use a relatively simple method (*see* **Note 1**) to purify megabase DNA from plant tissue that can be performed in any lab. Nuclei are isolated from plant tissue and purified nuclei are then processed using BioNano's Irys Prep Reagent kit to extract high molecular weight DNA. The DNA is subjected to nicking, labeling, repair, and staining before visualization on the IrysChip using microscopy.

## 2 Materials

### 2.1 Consumables and Equipment

1. Laboratory blender (1 liter size such as Waring seven speed blender).
2. Cheesecloth.
3. Miracloth.
4. CHEF Mammalian Genomic DNA Plug Kit (e.g., from Bio-Rad).
5. Clean-cut agarose (such as Bio-rad CleanCut™ Agarose).
6. Proteinase K.
7. RNaseA at 100 mg/ml or 7000 U/ml.
8. GELase™ (Epicentre, Madison, WI, USA) or equivalent.
9. Nt.BspQ1 10 U/μl.
10. Taq DNA ligase.
11. Taq DNA polymerase.
12. 10× ThermoPol® buffer for Taq DNA polymerase (New England Biolabs, Beverly, MA, USA) or equivalent.
13. β-Nicotinamide adenine dinucleotide (NAD+).
14. BioNano Genomics, San Diego, CA, USA: Irys machine, Irys Prep Reagent kit (custom) contains 10× buffer 3, 10× labeling buffer, 10× labeling mix, 50× repair mix, BioNano Stop solution, 4× flow buffer, 1 M DTT, DNA Stain, BioNano Lysis buffer.
15. Wide bore/orifice tips for p100 pipette.
16. Focused ultrasonicator.
17. Fluorometer and required assay kits/solutions.
18. Nitrocellulose membrane filters, 0.1 μm VCWP.

### 2.2 Buffers

1. **10× Homogenization buffer (HB)**: 0.02 M spermine tetrahydrochloride, 0.02 M spermidine trihydrochloride, 0.10 M EDTA, 0.10 M Tris base, 0.70 M KCl.

Add reagents and distilled water to 50–60% final volume and mix well. Bring to pH 9.4–9.5 with NaOH while mixing. Bring to final volume with distilled water and mix well. Solution does not need to be sterilized. It should *not* be autoclaved. Store at 4 °C for up to 1 year.

2. **1× HB + BME buffer**: 1× HB from 10× HB stock, 0.5 M sucrose, 0.4% β-mercaptoethanol (BME).

Add reagents and distilled water to 50–60% final volume and mix well. Bring to final volume with distilled water and mix well. BME is added to the buffer on the day of isolation in the fume hood. Solution should *not* be autoclaved. Store on ice until use.

3. **H + 20 buffer**: 1× HB from 10× HB stock, 0.5 M sucrose, 2-% Triton-X100.

Add 10× HB, sucrose, and distilled water to 50–60% final volume and mix well. When they are completely mixed, add the Triton-X100 slowly while stirring. Bring to final volume with distilled water and mix well. Solution should *not* be autoclaved. Store at 4 °C for up to 4 months.

4. **Nuclei wash buffer (NWB)**: 1× HB from 10× HB stock, 0.5 M sucrose, 1% Triton-X100, β-mercaptoethanol (BME).

Add 10× HB, sucrose, and distilled water to 50–60% final volume and mix well. When they are completely mixed, add the Triton-X100 slowly while stirring. Bring to final volume with distilled water and mix well. BME is added to the buffer on the day of isolation in the fume hood. Solution should *not* be autoclaved. Store on ice until use.

5. **TE buffer**: 10 mM Tris–HCl, pH 8.0, 1 mM EDTA, pH 8.0.

Add reagents and distilled water to 50–60% final volume and mix well. Bring to final volume with distilled water and mix well. Sterilize solution either through filtration or autoclave. Store at room temperature.

6. **1× Wash buffer**: 0.2 Tris base, 0.5 M EDTA.

Add reagents and distilled water to 50–60% final volume and mix well. Adjust pH to 8.0 using HCl. EDTA will not go into solution until the pH is 8.0. Bring to final volume with distilled water and mix well. Sterilize solution either by filtration or by autoclave. Store at room temperature.

7. **Cell suspension buffer (CSB)**: 10 mM Tris base pH 7.2, 50 mM EDTA, 20 mM NaCl.

Add reagents and distilled water to 50–60% final volume and mix well. Adjust pH to 7.2 using HCl. Bring to final volume with distilled water and mix well. Sterilize solution either by filtration or by autoclave. Store at room temperature.

8. **Proteinase K buffer**: 40 μl proteinase K (0.6 AU/μl), 500 μl BioNano Lysis buffer.

9. **RNase buffer** for each agarose plug sample tube: 2.5 ml TE buffer, 50 µl RNaseA (100 mg/ml or 7000 U/ml).

10. **β-Nicotinamide adenine dinucleotide (NAD+)**: Make a 50 mM NAD+ solution in water.

### 2.3 Safety

1. The Safety Data Sheet (SDS) for each chemical that is used in the protocol should be read and understood by the individual working with the protocol before the worker starts, particularly those areas pertaining to exposure controls (personal protective equipment and engineering controls), first aid measures, and accidental release measures. When working with multiple chemicals at once (i.e., solutions), individuals should follow the instructions on the SDS for the most hazardous chemical in the solution. Waste generated from this protocol should be discarded according to the recommendations of the SDS.

2. The manual for each piece of equipment that is used in the protocol should be read and understood by the individual working with the protocol before the worker starts, particularly those areas pertaining to exposure controls (personal protective equipment and engineering controls) and first aid measures.

3. All steps in which the individual is working with solutions that contain β-mercaptoethanol must use the appropriate personal protective equipment. This includes eye protection, hand protection, and body protection. Work with either concentrated β-mercaptoethanol or any solutions containing β-mercaptoethanol must take place in a fume hood.

## 3  Methods

### 3.1  Nuclei Isolation and Embedding

#### 3.1.1  Homogenize Leaf Tissue

1. Collect plant tissue (*see* **Note 2**), wrap in labeled paper towel, and place on ice. Plant tissue can be kept in ice for a few hours before processing.

2. Place labeled 250 ml centrifuge bottle(s) for each sample on ice. Add 5 ml of H + 20 buffer to each bottle and keep chilled on ice until use.

3. For each sample, weigh out approximately 5 g of plant tissue and keep on ice until ready to blend.

4. Cut leaf tissue for sample into approximately 1 in. pieces with scissors and place into laboratory blender (*see* **Note 2**—about cleaning scissors).

5. Add 150 ml of 1× HB + BME buffer to leaf tissue (*see* **Note 3**). Place lid on blender. Grind on middle speed setting (on a Waring brand blender, speed is 4) for 60 s.

6. Filter homogenate through a funnel lined with two layers of cheesecloth and one layer miracloth into the correspondingly

labeled 250 ml centrifuge bottle (*see* **Note 3**). With a glass rod, remove remaining leaf tissue debris from blender and place onto filter. Wring excess liquid from macerated leaf tissue down into funnel, being careful so the liquid does not squirt out the side.

7. Cap bottle and invert four times to mix.

8. Incubate on ice for a minimum of 30 min (maximum 3 h).

9. Repeat **steps 3–8** of Subheading 3.1.1 until all samples are processed.

10. While samples are incubating on ice, prepare nuclei wash buffer (NWB). Make enough NWB for five washes of 50 ml per sample. Place NWB on ice.

*3.1.2  Pellet and Rinse Nuclei*

1. Using scale, balance bottles containing homogenate samples with 1× HB + BME buffer. If needed, create a balance bottle with water.

2. Centrifuge samples at $3300 \times g$ for 20 min at 4 °C to pellet nuclei. After centrifugation, handle bottles with care so as not to dislodge the nuclei pellet.

3. Decant supernatant into BME/hazardous waste. Leave bottle upside down on clean paper towel for 30 s. Turn bottle upright and place back on ice.

4. Add 1 ml of chilled NWB buffer to the pellet (*see* **Note 2**). Gently resuspend pellet either through swishing the liquid around or using a clean paintbrush. It is important to break up any clumps that exist.

5. Add an additional 49 ml of chilled NWB to nuclei suspension. Gently swish the liquid around the bottom of tube.

6. Incubate on ice for 5 min.

7. Repeat **steps 2–6** of Subheading 3.1.2 for a minimum of three rinses for each sample. If necessary wash more than three times until pellet is pale green/beige in color (*see* **Note 4**).

8. After final rinse, add 1 ml of ice cold cell suspension buffer (CSB) to pellet.

9. Gently resuspend pellet either through swishing or using a clean paintbrush. Keep on ice until ready to proceed to next step.

*3.1.3  Embed Nuclei into Agarose Plugs*

1. Melt 2 % clean-cut agarose (such as Bio-rad CleanCut) with microwave oven (*see* **Note 5**).

2. Place agarose into a 43 °C water bath to guarantee uniformity in melting. Let the agarose equilibrate to 43 °C for a minimum of 5 min.

3. Equilibrate the agarose plug cast on ice for preparation of plugs. Make sure the plug cast is sealed on the bottom.

4. Incubate nuclei/CSB suspension from Subheading 3.1.2, **step 9** at 43 °C for 10 min in water bath.

5. Warm CSB to 43 °C for 10 min in water bath. This solution will be used for the nuclei suspension dilution series.

6. In labeled 2 ml tubes, prepare a nuclei suspension for a dilution series (Fig. 3). Keep all tubes at 43 °C until ready for **step 9**.

   (a) Tube A is undiluted, add 120 µl of nuclei from **step 4** of Subheading 3.1.3 using wide bore pipette tip (Fig. 3 step i). The final volume for tube A is 120 µl.

   (b) Tube B is diluted 1:2, add 800 µl of nuclei from **step 4** of Subheading 3.1.3 using wide bore pipette tip to 800 µl of equilibrated CSB (Fig. 3 step ii). The volume for tube B is 1600 µl.

   (c) Tube C has a final dilution of 1:4, add 800 µl from tube B using wide bore pipette tip to 800 µl of equilibrated CSB (Fig. 3 step iii). The volume for tube C is 1600 µl.

   (d) Tube D has a final dilution of 1:8, add 800 µl from tube C using wide bore pipette tip to 800 µl of equilibrated CSB (Fig. 3 step iv). The volume for tube D is 1600 µl.

   (e) Tube E has a final dilution of 1:16, add 800 µl from tube D using wide bore pipette tip to 800 µl of equilibrated CSB (Fig. 3 step v). The final volume for tube E is 1600 µl.

7. Mix each dilution by pipetting up and down with a wide bore tip three times before adding to the next tube, being careful not to introduce air bubbles. If air bubbles are introduced,

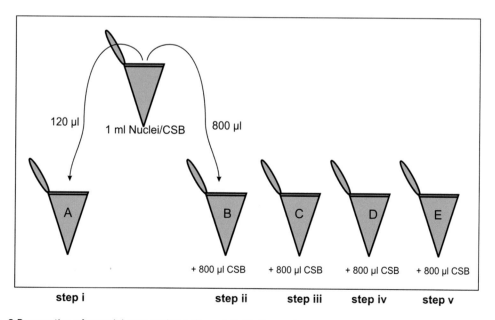

**Fig. 3** Preparation of a nuclei suspension using a 1:2 dilution series

leave the suspension at 43 °C for another 2 min allowing the bubbles to pop on their own.

8. Add the correct volume (Fig. 1) of equilibrated clean-cut agarose to tubes for the dilution series. In tube A of Fig. 1, add 120 µl of agarose.In tubes B, C, and D of Fig. 1, add 800 µl of agarose. In tube E of Fig. 1, add 1600 µl of agarose.

9. Mix by pipetting up and down with a wide bore tip (*see* **Note 6**) three times, being careful not to introduce air bubbles.

10. Pipette 95 µl of agarose-nuclei suspension into each well of the plug mold equilibrated on ice in **step 3** of Subheading 3.1.3. Incubate on ice to solidify for a minimum of 45 min.

*3.1.4  Proteinase K Treatment*

1. Remove plugs from molding cassette with the plunger from the kit and transfer to labeled 50 ml tube keeping each dilution separate (*see* **Note 7**).

2. Add 540 µl of proteinase K buffer to each sample tube.

3. Incubate for 120 min at 50 °C in a water bath. Make sure the plugs are in solution during the incubation.

4. Agitate every 15 min by gently inverting twice, make sure that the plugs remain in the solution when finished inverting. Alternatively, you can gently swirl instead of inverting (*see* **Note 8**).

5. Decant lysis solution using sieve adapter to retain the plug in the tube.

6. Add 540 µl of proteinase K buffer to each sample in tube.

7. Incubate overnight at 50 °C in a water bath.

8. Decant lysis solution using sieve adapter.

9. Add 10 ml of 1× Wash Buffer to the tube containing a plug.

10. Gently shake by hand for 10 s.

11. Decant wash buffer using sieve adapter. Repeat for a total of three washes.

12. Add 10 ml of 1× Wash Buffer to each tube.

13. Incubate on platform shaker at 150 RPM at room temperature for 15 min.

14. Decant the 1× Wash Buffer using sieve adapter. Repeat for a total of three washes.

15. Unless proceeding immediately to the RNaseA treatment, add 10 ml of 1× Wash Buffer and store at 4 °C.

*3.1.5  RNaseA Treatment*

1. Decant the 1× Wash Buffer using sieve adapter.

2. Add 10 ml of TE buffer to each tube, to equilibrate the plug(s).

3. Incubate samples on platform shaker at 150 RPM at room temperature for 15 min.

4. Decant TE buffer from tube using sieve adapter. Repeat for a total of two washes.

5. Add 2.5 ml of RNase buffer to each tube (no matter how many plugs).

6. Incubate at 37 °C for 60 min.

7. Decant RNaseA solution from tube using sieve adapter.

8. Add 10 ml of 1× Wash Buffer.

9. Incubate on platform shaker at 150 RPM at room temperature for 15 min.

10. Decant the wash buffer using sieve adapter. Repeat for a total of four washes.

11. Add 10 ml of 1× Wash Buffer and store plugs at 4 °C until ready to use (*see* **Note 9**).

**3.2 DNA Extraction and Cleanup**

*3.2.1 Melt Agarose Plug(s)*

1. Transfer plug(s) to labeled 50 ml tube. Only process plugs that you will be able to complete all the way through steps in Subheading 3.3 (the NLRS) reaction within 1 week.

2. Add 10 ml of TE buffer.

3. Incubate on platform shaker at 150 RPM at room temperature for 15 min.

4. Decant the TE buffer using sieve adapter. Repeat for a total of five washes.

5. Remove plug(s) from 50 ml tube with clean spatula.

6. Using fresh Kimwipe, gently blot plug to remove excess residual liquid from the TE buffer washes.

7. Transfer each plug to its corresponding labeled 1.5 ml microcentrifuge tube.

8. Centrifuge briefly to pool sample at bottom of tube.

9. Incubate microcentrifuge tubes containing plugs, at 70 °C for 2 min in water bath to melt agarose.

10. Immediately transfer the tubes to a 43 °C water bath, and incubate for 5 min.

11. Add 2 μl of GELase. Gently mix by stirring three to five times with pipette tip.

12. Incubate tubes at 43 °C for 45 min.

*3.2.2 Clean Up DNA via Drop Dialysis*

1. Equilibrate dialysis membranes for a minimum of 10 min by floating it on top of 30 ml of TE buffer in a petri dish.

2. Add sample to the center of membrane using p200 pipette with a wide bore tip.

3. Cover petri dish and incubate at room temperature for 60 min allowing for the separation of digested agarose from the high molecular weight DNA.

4. Remove the sample from membrane and transfer it to a labeled 2 ml tube using a p200 pipette with a wide bore tip.

5. Pipette-mix sample five to six times very slowly, being careful to avoid introducing bubbles or shearing of DNA.

6. Incubate at room temperature overnight to homogenize.

7. Transfer sample to a 1.5 ml microcentrifuge tube.

*3.2.3  Quantify DNA*

1. Pipette-mix the viscous DNA three times using p200 pipette equipped with wide bore tip. Pipette slowly to avoid introducing air bubbles.

2. Transfer 4 μl from top, middle, and bottom fractions of sample to labeled fluorometer tube using a p20 pipette. Change tips between fractions.

3. Centrifuge sample briefly to collect the aliquot at bottom of fluorometer tube.

4. Sonicate/shear sample DNA using a sonicator (*see* **Note 10**).

5. Prepare samples for fluorometer analysis by adding detection dye according to manufacturer's recommendations (*see* **Note 11**).

6. Quantify DNA following fluorometer's protocol and record values.

7. DNA concentration from quantification should be 35 ng/μl $\leq [gDNA_{average}] \leq 200$ ng/μl. If it is not, you will have to repeat the experiment. The coefficient of variance percentage (CV%) between the top, middle, and bottom fractions should be less than 25 %; if it is not, you can let the sample sit at room temperature for additional equilibration (*see* **Note 12**).

**3.3  Nick, Label, Repair, and Stain (NLRS) Reactions (See Note 13)**

*3.3.1  Nick DNA*

1. Determine volume of genomic DNA (gDNA) required for either a 300 ng or a 900 ng reaction.

2. Prepare nicking reaction mix according to Table 1.

3. Aliquot reaction mixture without DNA and mix by pipetting up and down several times.

4. Add the genomic DNA to the nicking reaction mix in 200 μl PCR tube using a wide bore pipette tip. Using pipette set to three-fourths the volume of your reaction, with a wide bore tip, pipette-mix sample four times slowly. Centrifuge briefly.

5. Incubate at 37 °C for 120 min in a thermocycler.

6. Centrifuge briefly.

**3.3.2 Label DNA**

1. Prepare labeling reaction mix according to Table 2.

2. Add labeling reaction mix to nicked DNA solution from Subheading 3.3.1. Using a pipette set to half the volume of your reaction, with a wide bore tip, pipette-mix sample four times slowly. Centrifuge briefly.

3. Incubate at 72 °C for 60 min in a thermocycler.

4. Centrifuge briefly.

**3.3.3 Repair DNA**

1. Prepare repairing reaction mix according to Table 3.

2. Add repairing reaction mix to the nick-labeled DNA solution. Using pipette set to half the volume of your reaction, with a wide bore tip, pipette-mix sample four times slowly. Centrifuge briefly.

3. Incubate at 37 °C for 30 min in a thermocycler.

4. Centrifuge briefly.

5. Add 1 µl of BioNano Stop Solution for the 300 ng reaction or 3 µl of BioNano Stop Solution for the 900 ng reaction. Gently mix by stirring five times with pipette tip. Centrifuge briefly.

**3.3.4 Stain DNA**

1. Prepare staining reaction mix according to Table 4.

2. Add 120 µl of staining reaction mix to the nick-label-repair (NLR) DNA solution. Centrifuge briefly.

3. Transfer the entire nick-label-repair with stain (NLRS) DNA solution to a labeled 0.5 ml amber tube.

4. Pipette-mix the sample three times using p200 pipette equipped with a wide bore tip set to 150 µl volume. Pipette slowly to avoid introducing bubbles. Incubate at 4 °C overnight.

5. The NLRS DNA is good for 6 weeks. Store at 4 °C protected from light.

**Table 1**
**Nicking reaction**

| | 1× volume | |
| --- | --- | --- |
| | 300 ng | 900 ng |
| gDNA | __µl* | __µl* |
| 10× buffer 3 (BioNano) | 1.0 µl | 3.0 µl |
| 10 U/µl Nt.BspQ1 | 0.9 µl | 2.7 µl |
| Ultrapure H₂O | (8.1-__µl gDNA) | (24.3-__µl gDNA) |
| Final volume | 10 µl | 30 µl |

* Volume of DNA is to be calculated by the user based on the concentration of their particular DNA sample and a final amount of either 300 ng or 900 ng DNA in the reaction.

**Table 2**
**Labeling reaction**

|  | 1× volume | |
| --- | --- | --- |
|  | 300 ng | 900 ng |
| Nicked DNA | 10 µl | 30 µl |
| 10× labeling buffer (BioNano) | 1.5 µl | 4.5 µl |
| 10× labeling mix (BioNano) | 1.5 µl | 4.5 µl |
| 5 U/µl Taq polymerase | 1.0 µl | 3.0 µl |
| Ultrapure $H_2O$ | 1.0 µl | 3.0 µl |
| Final volume | 15 µl | 45 µl |

**Table 3**
**Repairing reaction**

|  | 1× volume | |
| --- | --- | --- |
|  | 300 ng | 900 ng |
| Nicked-labeled DNA | 15 µl | 45 µl |
| 10× thermo pol buffer | 0.5 µl | 1.5 µl |
| 50× repair mix (BioNano) | 0.4 µl | 1.2 µl |
| 50 mM NAD+ | 0.4 µl | 1.2 µl |
| Taq DNA ligase | 1.0 µl | 3.0 µ |
| Ultrapure $H_2O$ | 2.7 µl | 8.1 µl |
| Final volume | 20 µl | 60 µl |

**Table 4**
**Staining reaction**

|  | 1× volume | |
| --- | --- | --- |
|  | 300 ng | 900 ng |
| Nicked-labeled-repaired DNA | 20 µl | 60 µl |
| 4× flow buffer (BioNano) | 15 µl | 45 µl |
| 1 M DTT (BioNano) | 12 µl | 36 µl |
| DNA stain (BioNano) | 1.5 µl | 4.5 µl |
| Ultrapure $H_2O$ | 11.5 µl | 34.5 µl |
| Final volume | 60 µl | 180 µl |

**3.3.5   Quantify NLRS DNA**

1. Remove the NLRS DNA from 4 °C and warm to room temperature for a minimum 30 min.

2. Pipette-mix NLRS DNA three times using a p200 pipette equipped with a wide bore tip set to 150 µl volume slowly. Centrifuge briefly.

3. Transfer 4 µl from the top, middle, and bottom fractions of sample to labeled fluorometer tubes using a p20 pipette with standard tips. Change tips between fractions. Centrifuge sample briefly to collect sample at bottom of the tube.

4. Sonicate/shear sample DNA using a sonicator (*see* Subheading 3.2.3 and **Notes 10, 11**, and **12**). Prepare fluorometer samples according to manufacturer's recommendations.

5. Quantify DNA following fluorometer's protocol and record values. Before loading sample onto the Irys chip, confirm that the DNA concentration is within 3 ng/µl $\leq$ [NLRS DNA$_{average}$] $\leq$ 10 ng/µl and coefficient of variance percentage (CV%) between sample fractions is below 25 %. If CV% is greater than 25 %, let the samples equilibrate at room temperature for a day. If the CV% is still too high, you will need to start over.

**3.4   Run Sample on Irys**

Load sample(s) on to Irys according to the manufacturer's instructions.

**3.4.1   Loading and Running Irys**

# 4   Notes

1. There are many ways to purify megabase DNA from plants, most of which depend on the liberation of intact nuclei or protoplasts from plant tissue followed by one of several methods to purify the nuclei/protoplasts before embedding in agarose plugs for lysis and protein removal. The key to purification of megabase DNA from plant tissue is to purify intact chromosomes encapsulated or fixed (i.e., not elongated in solution) from insoluble material such as fiber, starch, and cell wall before complete solubilization of DNA, washing, and protein removal. Methods for liberation of nuclei include grinding tissue in liquid nitrogen and cutting fresh tissue in a blender or with a razor blade, followed by filtering large insoluble material, washing remaining material, and then, sometimes, further separation by Ficoll gradients or cushions and washing nuclei before embedding in agarose gel plugs (*see* ref. 3, 4). Another method that may be used to produce very high-quality DNA involves purification of nuclei or metaphase chromosomes by

flow sorting (*see* ref. 5). In other methods, protoplasts can be prepared by enzymatic removal of the cell wall followed by embedding protoplasts in gel plugs (*see* ref. 6). In the current protocol, we use a relatively simple method that can be performed in any lab. When more involved methods are used, it may be possible to produce higher-quality DNA, which, in turn, results in longer single molecules and lower error rates in the Irys system (unpublished data BioNano Genomics).

2. Plant tissue must be collected the morning of the DNA isolation and kept on ice until ready to use. Do not use older or yellow tissue. Clean scissors before use to avoid contamination.

3. All steps with solutions that contain β-mercaptoethanol must be performed in a fume hood. Place used cheesecloth, miracloth, paper towels, gloves, and leaf tissue into a properly labeled hazardous waste bag prepared for this isolation if they have had contact with β-mercaptoethanol.

4. The more nuclei rinses performed, the cleaner the final DNA will be, although with each wash the overall DNA yield is reduced.

5. Be careful not to burn yourself or to overboil the liquid. Repeated short bursts of microwave (i.e., 10 s) work best.

6. Wide bore tips decrease the shearing of large DNA molecules during pipetting. It is critical to use wide bore tips and handle samples gently to assure large DNA molecules for downstream analysis.

7. Plug molds can be rinsed in distilled water, treated with 1 M hydrochloric acid overnight, and rinsed again in sterile distilled water, dried, and stored for later use. Do not autoclave the plug molds.

8. Do not use a hybridization oven as this tends to leave the plugs on the side of the tube and out of solution.

9. Plugs can be stored for up to 6 months at 4 °C in 1× Wash Buffer, but stability will depend on the purity of the DNA.

10. We use a Covaris sonicator with the following parameters: treatment time 60s, number of cycles 5, duty cycle 5%, intensity 3, and cycles/burst 500. If a sonicator is not available, DNA should be sheared to approximately 1000 bp.

11. We use a Qubit from ThermoFisher Scientific. Spin down samples after sonication. Add 196 μl prepared Qubit BR ds DNA Assay kit buffer-dye solution. Vortex for 3–5 s. Centrifuge briefly. Let stand at room temperature for 2 min before DNA quantification.

12. If samples are consistently nonhomogeneous, you can skip Subheading 3.2.2 **step 6**, but your DNA yield will be lower.

13. All NLRS reaction mixes should be created with 0.5× overage.

## Acknowledgement

This work was supported by the National Science Foundation grant IOS-1238231.

## References

1. Schwartz DC, Li X, Hernandez LI, Ramnarain SP, Huff EJ, Wang YK (1993) Ordered restriction maps of Saccharomyces cerevisiae chromosomes constructed by optical mapping. Science 262(5130):110–114

2. Luo M-C, Gu YQ, You FM et al (2013) A 4-gigabase physical map unlocks the structure and evolution of the complex genome of Aegilops tauschii, the wheat D-genome progenitor. Proc Natl Acad Sci U S A 110:7940–7945

3. Dvorak J, McGuire PE, Cassidy B (1988) Apparent sources of the A genomes of wheats inferred from the polymorphism in abundance and restriction fragment length of repeated nucleotide sequences. Genome 30:680–689

4. Zhang M, Zhang Y, Scheuring CF, Wu C-C, Dong JJ, Zhang H-B (2012) Preparation of megabase-sized DNA from a variety of organisms using the nuclei method for advanced genomics research. Nat Protoc 7(3):467–478

5. Simkova H, Cihalikova J, Vrana J, Lysak MA, Dolezel J (2003) Preparation of HMW DNA from plant nuclei and chromosomes isolated from root tips. Biologia Plantarum 46(3): 369–373

6. van Daelen RA, Jonkers JJ, Zabel P (1989) Preparation of megabase-sized tomato DNA and separation of large restriction fragments by field inversion gel electrophoresis (FIGE). Plant Mol Biol 12:341–352

# Chapter 10

## Flow Sorting Plant Chromosomes

**Jan Vrána, Petr Cápal, Jarmila Číhalíková, Marie Kubaláková, and Jaroslav Doležel**

### Abstract

Nuclear genomes of many important plant species are tremendously complicated to map and sequence. The ability to isolate single chromosomes, which represent small units of nuclear genome, is priceless in many areas of plant research including cytogenetics, genomics, and proteomics. Flow cytometry is the only technique which can provide large quantities of pure chromosome fractions suitable for downstream applications including physical mapping, preparation of chromosome-specific BAC libraries, sequencing, and optical mapping. Here, we describe step-by-step procedure of preparation of liquid suspensions of intact mitotic metaphase chromosomes and their flow cytometric analysis and sorting.

**Key words** Cell cycle synchronization, Chromosome isolation, Cytogenetic stocks, FISH, FISHIS, Flow cytometry and sorting, Metaphase accumulation, Plants

## 1 Introduction

Three decades of flow cytometric analysis and sorting of plant chromosomes have had a significant impact on several fields of plant research, especially genomics (reviewed in [1]). The advantage of chromosome-based approach in genomics is that it simplifies the analysis of large and complex plant genomes by working with smaller parts, reducing the sample complexity up to more than one order of magnitude. It all started in 1984, inspired by successful flow cytometric sorting of human and animal chromosomes [2, 3], when the first plant species, *Haplopappus gracilis*, was used for chromosome analysis, using then relatively new technology [4]. Since this time, many improvements to the technique were made, and to date, chromosome analysis and sorting is reported for 24 plant species from 18 genera [1]. Nevertheless, the progress of plant chromosome flow cytometry is hindered by several specific issues (reviewed in [5]): (a) insufficient amount of actively dividing cells in intact plant tissues and lack of sufficient

Shahryar F. Kianian and Penny M.A. Kianian (eds.), *Plant Cytogenetics: Methods and Protocols*,
Methods in Molecular Biology, vol. 1429, DOI 10.1007/978-1-4939-3622-9_10,
© Springer Science+Business Media New York 2016

cell cycle synchrony, (b) rigid cell walls hampering the release of chromosomes from plant cells, and (c) inability to discriminate each chromosome type due to similarities in DNA content among chromosomes. The problems with preparation of suspensions of intact metaphase chromosomes were overcome in 1992 by Doležel et al. [6] who developed an isolation method based on synchronization of root tip meristem cells and subsequent release of metaphase chromosomes from formaldehyde-fixed root tips using mechanical homogenization. As this method proved to be simple and reproducible, other plant species followed suit soon after [7–11]. But from the very beginning, it was obvious that not all chromosomes could be resolved from each other, and this would limit the development of plant flow cytogenetics and subsequent applications of sorted chromosomes. Unlike the situation in animal and human flow cytogenetics, simultaneous staining of chromosomes with two different DNA base pair-specific dyes did not help in discriminating more chromosomes [12, 13]. Thus, at present, the two best methods for discrimination of different plant chromosome types are (a) using cytogenetic stocks (e.g., translocation, deletion, or alien addition chromosome lines; [14]) and (b) fluorescent labeling of specific DNA sequences using fluorescence in situ hybridization in suspension (FISHIS; [15]). As FISHIS is relatively a new method, it has not had much impact on plant chromosome genomics. Moreover, it seems this method is not compatible with all types of DNA probes, which could limit its wider use (personal observation). On the other hand, since the authors introduction to plant flow cytogenetics [16], cytogenetic stocks helped in many ways, such as obtaining otherwise unsortable chromosomes [17, 18], mapping of genes to subchromosomal regions [19, 20], delimiting the translocations [21], and mapping alien introgressions [22], among others. Moreover, flow-sorted chromosome arms from wheat ditelosomic lines were chosen by The International Wheat Genome Sequencing Consortium (IWGSC) as a primary template for sequencing the huge genome of hexaploid wheat [23–25]. To conclude, applications of flow-sorted plant chromosomes, collectively termed chromosome genomics, are wide-ranging and include physical mapping-specific DNA sequences using either FISH [26, 27] or PCR [19, 28, 29], amplification of chromosomal DNA using multiple displacement amplification (MDA) [30, 31] and subsequent sequencing using NGS technologies [32–36], preparation of high molecular weight (HMW) DNA [37], integration of genetic and physical maps [38], development of DNA markers for positional cloning [39–41], and creation of chromosome-specific BAC libraries [42–44]. In the near future, the expansion of new applications of sorted chromosomes such as single-chromosome sequencing (Cápal et al., in preparation), optical mapping (H. Staňková and Z. Milec, personal communication), and analysis of protein composition of mitotic chromosomes [45] will progress.

This protocol describes the preparation of liquid suspensions of intact mitotic metaphase chromosomes and their flow cytometric analysis through sorting for use in various downstream applications.

# 2    Materials

### 2.1    Plant Material

Dried viable and vernalized seeds of *Aegilops* (goat grasses, and different subspecies), *Avena sativa* (oat), *Cicer arietinum* (chickpea), *Hordeum vulgare* (barley), *Pisum sativum* (pea), *Secale cereale* (rye), *Silene latifolia* (white campion), *Triticum aestivum* (bread wheat), *Triticum durum* (durum wheat), *Vicia faba* (faba bean), and *Vicia sativa* (common vetch).

### 2.2    Reagents and Solutions

#### 2.2.1    Reagents and Solutions for Cell Cycle Synchronization, Accumulation of Metaphases, Preparation of Chromosome Suspensions, and Chromosome Sorting

1. Solution A: 45 mM $H_3BO_3$ (280 mg), 20 mM $MnSO_4 \cdot H_2O$ (340 mg), 0.4 mM $CuSO_4 \cdot 5H_2O$ (10 mg), 0.8 mM $ZnSO_4 \cdot 7H_2O$ (22 mg), and 0.08 mM $(NH_4)_6Mo_7O_{24} \cdot 4H_2O$ (10 mg) in deionized water (100 mL). Store at 4 °C.

2. Solution B: 0.05 mM concentrated $H_2SO_4$ (0.5 mL) in deionized water (100 mL). Store at 4 °C.

3. Solution C: 18 mM $Na_2EDTA$ (3.36 g) and 20 mM 2.79 g $FeSO_4$ (20 mM) in deionized water. Heat the solution to 70 °C while stirring until the color turns yellow brown. Cool down, adjust the volume with deionized water (500 mL), and store at 4 °C.

4. Hoagland's stock solution (10×): 4.7 g $Ca(NO_3)_2 \cdot 4H_2O$ (40 mM), 2.6 g $MgSO_4 \cdot 7H_2O$ (20 mM), 3.3 g $KNO_3$ (65 mM), 0.6 g $NH_4H_2PO_4$ (10 mM), 5 mL solution A, and 0.5 mL solution B, in deionized water. Adjust volume to 500 mL. Prepare just before use.

5. Hoagland's nutrient solution (1×): 100 mL Hoagland's stock solution (10×) and 5 mL solution C in deionized water. Adjust volume to 1000 mL. Prepare just before use.

6. Hoagland's nutrient solution (0.1×): 10 mL Hoagland's stock solution (10×) and 0.5 mL solution C in deionized water. Adjust volume to 1000 mL. Prepare just before use.

7. 1 mM hydroxyurea (HU) solution: dissolve 60.8 mg hydroxyurea in 800 mL 0.1× Hoagland's nutrient solution. Prepare just before use.

8. 1.25 mM HU solution: dissolve 76 mg hydroxyurea in 800 mL 0.1× Hoagland's nutrient solution. In case of faba bean and chickpea, use 1× Hoagland's nutrient solution instead. Prepare just before use.

9. 2 mM HU solution: dissolve 121.6 mg hydroxyurea in 800 mL 0.1× Hoagland's nutrient solution. Prepare just before use.

10. 2.5 mM HU solution: dissolve 152 mg hydroxyurea in 800 mL 0.1× Hoagland's nutrient solution. Prepare just before use.

11. Amiprophos methyl (APM) stock solution (20 mM): dissolve 60.86 mg APM in 10 mL ice-cold acetone and store at −20 °C in 1 mL aliquots.

12. APM working solution (2.5 μM): 101.3 μL APM stock solution in 800 mL deionized water. Prepare just before use.

13. APM working solution (10 μM): 405.2 μL APM stock solution in 800 mL deionized water. Prepare just before use.

14. Oryzalin stock solution (10 mM): dissolve 86.59 mg oryzalin in 25 mL ice-cold acetone. Store at −20 °C in 1 mL aliquots.

15. Oryzalin working solution (2.5 μM): 200 μL oryzalin stock solution in 800 mL deionized water. Prepare just before use.

16. Oryzalin working solution (5 μM): 400 μL oryzalin stock solution in 800 mL deionized water. Prepare just before use.

17. Oryzalin working solution (10 μM): 800 μL oryzalin stock solution in 800 mL deionized water. Prepare just before use.

18. Tris buffer: 10 mM Tris (606 mg), 10 mM $Na_2EDTA$ (1.861 g), 100 mM NaCl (2.922 g) in deionized water (500 mL). Adjust pH to 7.5 using 1 N NaOH.

19. Formaldehyde 2 % fixative: 13.5 mL formaldehyde in Tris buffer. Adjust volume to 250 mL. Prepare just before use.

20. Formaldehyde 4 % fixative: 27 mL formaldehyde in Tris buffer. Adjust volume to 250 mL. Prepare just before use.

21. LB01 buffer: 15 mM Tris (0.363 g), 2 mM $Na_2EDTA$ (0.149 g), 0.5 mM spermine·4HCl (0.0348 g), 80 mM KCl (1.193 g), 20 mM NaCl (0.234 g), 0.1 % (v/v) Triton X-100 (200 μL) in deionized water (200 mL). Adjust pH to 9. Filter through a 0.22 μm filter to remove small particles. Add 220 μL β-mercaptoethanol and mix well. Store at −20 °C in 8 mL aliquots.

22. 10× HKS stock solution: 100 mM Tris (0.606 g), 100 mM $Na_2EDTA$ (1.86 g), 10 mM spermine·4HCl (0.174 g), 10 mM spermidine·3HCl (0.127 g), 1.3 M KCl (4.84 g), 200 mM NaCl (0.585 g), 1 % (v/v) Triton X-100 (500 μL) in deionized water (50 mL). Adjust pH to 9.4. Filter through a 0.22 μm filter to remove small particles. Store at 4 °C.

23. HKS buffer: 1.5 mL 10× HKS stock solution in 8.5 mL deionized water. Add 11 μL β-mercaptoethanol. Prepare just before use.

24. 10× LB01-P stock solution: 15 mM Tris (181.5 mg), 2 mM $Na_2EDTA$ (74.4 mg), 0.5 mM EGTA (19 mg), 0.2 mM spermine·4HCl (6.96 mg), 0.5 mM spermidine·3HCl (12.73 mg), 80 mM KCl (596.5 mg), 20 mM NaCl (116.9 mg),

0.1% (v/v) Triton X-100 (100 μL) in deionized water (100 mL). Adjust pH to 9. Filter through a 0.22 μm filter to remove small particles. Store at –20 °C, in 10 mL aliquots.

25. LB01-P buffer: Thaw the 10× stock solution at room temperature. Mix 1 mL 10× stock solution in 9 mL deionized water. Add 10 μL β-mercaptoethanol and mix well. Prepare just before use.

26. 4′,6-Diamidino-2-phenylindole (DAPI) stock solution (0.1 mg/mL): dissolve DAPI in deionized water by stirring. Filter through a 0.22 μm filter to remove small particles. Store at –20 °C, in 0.5 mL aliquots.

27. Phenylmethanesulfonylfluoride (PMSF) stock solution (100 mM): dissolve PMSF in isopropanol. Store at –20 °C in 100 μL aliquots.

28. Reaction mix for single-chromosome sorting: mix 30 μL of sample buffer from whole genome amplification kit (such as the GenomiPhi V2 kit, GE Healthcare) and 3 μL proteinase K solution (RNase and DNase free, 10 mg/μL). Briefly vortex. Prepare just before use.

*2.2.2  Reagents and Solutions for FISH*

1. P5 buffer: 10 mM Tris (30.28 mg), 50 mM KCl (93.2 mg), 2 mM $MgCl_2 \cdot 6H_2O$ (10.17 mg), and 5% sucrose (1.25 g) in deionized $H_2O$ (25 mL). Adjust pH to 8 using 1 N HCl. Store at –20 °C in 1 mL aliquots.

2. 20× SSC stock solution: 3 M NaCl (175.3 g) and 300 mM $Na_3C_6H_5O_7 \cdot 2H_2O$ (88.2 g) in deionized $H_2O$ (1000 mL). Adjust pH to 7. Sterilize by autoclaving. Store at room temperature.

3. 4× SSC washing buffer: 20× SSC (200 mL) and 0.2% Tween 20 in deionized $H_2O$.

4. 2× SSC washing buffer: 20× SSC (100 mL) in deionized $H_2O$. Prepare just before use.

5. 0.1× SSC stringent washing buffer: 20× SSC (5 mL), 0.1% Tween 20, and 2 mM $MgCl_2 \cdot 6H_2O$ in deionized $H_2O$. Prepare just before use.

6. Hybridization mix: 40% formamide (10 μL), 20× SSC (1.25 μL), 0.625 μL calf thymus (250 ng/μL), labeled DNA probe(s) (1 ng/μL). Add 50% dextran sulfate (final volume 25 μL). Prepare just before use. Labeled DNA probes (either directly labeled with fluorescent probes or labeled by digoxigenin or biotin) may be prepared using PCR [31].

7. Detection of digoxigenin-labeled probes: FITC-labeled anti-digoxigenin antibody raised in sheep.

8. Detection of biotin-labeled probes: Cy3-labeled streptavidin antibody.

9. Blocking solution: dissolve 0.5 g blocking reagent in 50 mL 4× SSC. Autoclave. Store at –20 °C in 1 mL aliquots.

10. Vectashield antifade solution containing DAPI.

*2.2.3 Reagents and Solutions for FISHIS*

1. 10 M NaOH: dissolve solid NaOH in deionized water. Store at room temperature.

2. 1 M Tris–HCl: dissolve Tris in deionized water by stirring; adjust the pH to 7.5 using 1 N HCl. Store at 4 °C.

3. (GAA)$_7$ microsatellite probe labeled with FITC: dissolve the probe to 100 μM concentration with 2× SSC according to manufacturer's instructions. Prepare working solution by adding 2× SSC to final concentration 80 ng/μL. Store in the dark at –20 °C in 20 μL aliquots.

*2.3 Instruments and Other Utilities*

1. Biological incubator with temperature control.

2. Glass petri dishes (18 cm diameter) with filter paper cut to 18 cm diameter or plastic trays (4000 mL) with perlite for seed germination.

3. Plastic box (750 mL) including plastic cover with drilled holes (1–3 mm in diameter).

4. Aquarium aerating system with air stones.

5. Mechanical tissue homogenizer (e.g., Polytron PT 1300D, Kinematica AG, Littau, Switzerland) or sharp razor blade.

6. Nylon mesh filters (20 and 50 μm pore size, respectively), cut to 4×4 cm squares.

7. Sample tubes for flow cytometer.

8. pH meter.

9. Flow cytometer and sorter equipped with blue (488 nm) and UV (355 nm) or violet (405 nm) lasers.

10. Microscopic slides with coverslips.

11. Fluorescence microscope with optical filter sets for DAPI, FITC, Cy3, and Texas Red fluorescence.

12. Humidity chamber (temperature-controlled chamber containing wet tissues).

13. Rubber cement.

# 3  Methods

*3.1 Seed Germination (See Note 1)*

1. Leave the seeds to swell in a beaker filled with deionized H$_2$O. In case of small seeds (e.g., cereals, grasses), let them to soak for about 15 min at room temperature; in case of bigger seeds (e.g., legumes), keep them aerated overnight in the dark at 25 °C.

2. Germinate the seeds in a glass petri plate on a layer of wet paper towel sandwiched by two layers of filter paper in the dark at 25 °C, until optimal root length is achieved (typically 1–3 cm). In case of faba bean, put the seeds into a plastic tray containing wet perlite, and incubate in the dark at 25 °C.

3. Position the seedlings onto a plastic cover by threading the roots through the holes and put the cover onto plastic box filled with deionized water. Adjust the volume of water so that the root tips are fully immersed. At this point, the seedlings are ready for subsequent treatments.

*3.2 Cell Cycle Synchronization Using HU*

1. Transfer the plastic cover with seedlings onto a plastic box with HU solution and incubate by aerating in the dark at 25 °C. For appropriate HU concentrations and incubation times, *see* Table 1 (*see* **Note 2**).

2. Remove the seedlings from the HU solution and transfer them onto a plastic box filled with 0.1× Hoagland's solution (1× Hoagland's solution in case of faba bean and chickpea) and incubate by aerating in the dark at 25 °C. Check the recovery times for each plant species in the Table 1.

*3.3 Metaphase Accumulation (See Note 3)*

1. Transfer the cover with seedlings onto a box filled with solution of mitotic spindle inhibitor. For appropriate agent and its concentration and incubation time, *see* Table 2. The treatment is performed in the dark at 25 °C by aerating.

**Table 1**
**Cell cycle synchronization using hydroxyurea**

| Species | HU conc. (mM) | HU incubation time (h) | Recovery time (h) | References |
|---------|---------------|------------------------|-------------------|------------|
| *Aegilops* ssp. | 1.25 | 18 | 5 | [46] |
| *Avena sativa* | 2.0 | 18 | 4.5 | Not published |
| *Cicer arietinum* | 1.25 | 18 | 4 | [29], modified |
| *Hordeum vulgare* | 2.0 | 18 | 6.5 | [9], modified |
| *Pisum sativum* | 1.25 | 18 | 4.5 | [7], modified |
| *Secale cereale* | 2.5 | 18 | 6.5 | [11] |
| *Silene latifolia* | 2.0 | 18 | 5 | [47] |
| *Triticum aestivum* | 2.0[a] | 18 | 5.5 | [10], modified |
| *Triticum durum* | 1.25 | 18 | 5 | [48] |
| *Vicia faba* | 1.25 | 18.5 | 4.5 | [6], modified |
| *Vicia sativa* | 2.5 | 18.5 | 3.5 | [49], modified |

[a]In case of ditelosomic lines, use 1.25 mM HU (*see* **Note 2**)

**Table 2**
**Metaphase accumulation using mitotic spindle poison agent**

| Species | APM conc. | Oryzalin conc. | Incubation time | References |
|---|---|---|---|---|
| *Aegilops* ssp. | 2.5 µM | – | 2 h + ice overnight | [46] |
| *Avena sativa* | – | 10 µM | 2 h + ice overnight | Not published |
| *Cicer arietinum* | – | 5 µM | 2 h + ice overnight | [29], modified |
| *Hordeum vulgare* | 2.5 µM | – | 2 h + ice overnight | [9] |
| *Pisum sativum* | 10 µM | – | 2 h + ice overnight | [7], modified |
| *Secale cereale* | – | 10 µM | 2 h + ice overnight | [11] |
| *Silene latifolia* | – | 2.5 µM | 2 h | [47] |
| *Triticum aestivum* | 2.5 µM | – | 2 h + ice overnight | [10] |
| *Triticum durum* | 2.5 µM | – | 2 h + ice overnight | [48] |
| *Vicia faba* | 2.5 µM | – | 2 h | [6], modified |
| *Vicia sativa* | – | 5 µM | 2 h | [49] |

2. Rinse the roots in a container filled with deionized water.

3. Optional: remove the cover with seedlings and put it into a container filled with ice water containing ice cubes and keep it overnight in the refrigerator (*see* **Note 4**). Make sure that all roots are immersed. For the list of species and the length of treatment, *see* Table 2.

**3.4 Preparation of Liquid Chromosome Suspensions (See Note 5)**

1. Excise the roots (approx. 1 cm long) and transfer them into beaker with deionized $H_2O$. The number of roots necessary for preparation of 1 mL sample depends on species (Table 3).

2. Place the roots into the beaker containing formaldehyde fixative solution and incubate at 5 °C. Times and concentrations are given in Table 3.

3. Rinse the roots three times in Tris buffer at 5 °C for 5 min. Keep the roots in the Tris buffer on ice after the last wash.

4. Cut the root apices (1–2 mm long) and transfer them into a 5 mL polystyrene tube containing 1 mL LB01 buffer (*see* **Note 6**). Grind the root tips using a mechanical homogenizer. Settings for each species are specified in Table 3.

   Optional (*see* **Note 7**): chop the root apices in a drop of LB01 buffer using sharp razor blade until milky consistency is achieved. Add 1 mL LB01 buffer and transfer to 5 mL polystyrene tube.

5. Filter the crude suspension through 50 µm nylon mesh into new 5 mL polystyrene tube.

6. Keep the suspension on ice until flow cytometric analysis.

**Table 3**
**Preparation of chromosome suspensions**

| Species | No. of roots | Formaldehyde fixation (conc./duration) | Suspension preparation (RPM/duration) | References |
|---------|--------------|----------------------------------------|---------------------------------------|------------|
| *Aegilops* ssp. | 60 | 2%/20 min | 20,000/13 s | [46], modified |
| *Avena sativa* | 60 | 2%/25 min | 25,000/13 s | Not published |
| *Cicer arietinum* | 20 | 2%/30 min | 13,000/18 s | [29], modified |
| *Hordeum vulgare* | 60 | 2%/20 min | 15,000/13 s | [9], modified |
| *Pisum sativum* | 20 | 2%/20 min | 13,000/18 s | [7], modified |
| *Secale cereale* | 60 | 2%/30 min | 15,000/13 s | [11], modified |
| *Silene latifolia* | 180 | 2%/15 min | 18,000/13 s | [47], modified |
| *Triticum aestivum* | 60 | 2%/20 min | 20,000/13 s | [10], modified |
| *Triticum durum* | 60 | 2%/20 min | 20,000/13 s | [48], modified |
| *Vicia faba* | 20 | 4%/30 min | chopping | [6], modified |
| *Vicia sativa* | 20 | 2%/25 min | 13,000/18 s | [49], modified |

***3.5 Chromosome Labeling Using FISH in Suspension (FISHIS)***

1. Filter 300 µL of chromosome suspension (*see* **Note 8**) through 20 µm nylon mesh into 1.5 mL tube and place the tube on ice.

2. Add 10 M NaOH to reach a pH 13.

3. Incubate the suspension for 20 min on ice.

4. Adjust the pH in the range of 8–9 using Tris–HCl (*see* **Note 9**).

5. Immediately add 1 µL of $(GAA)_7$ probe working solution and let the suspension incubate for 1 h in the dark at room temperature.

6. Keep the suspension on ice until the flow cytometric analysis (*see* **Note 10**).

***3.6 Chromosome Sorting Using Flow Cytometry***

1. Start up and set up the flow sorter according to manufacturer's instructions.

2. Pass the suspension through the 20 µm nylon mesh.

3. Stain the suspension with DAPI to a final concentration 2 µg/mL (for 1 mL sample, use 20 µL of DAPI stock solution).

4. In acquisition software of flow sorter, open or create the appropriate histograms and dot plots. First, use dot plot FSC vs. DAPI for showing populations representing chromosomes (Fig. 1a). Create a region surrounding the chromosomes (R1) and use this gating on the remaining dot plots and histograms. For sorting, create dot plots DAPI-W vs. DAPI-A (Fig. 1b) in case of monoparametric experiment (DNA staining only) or

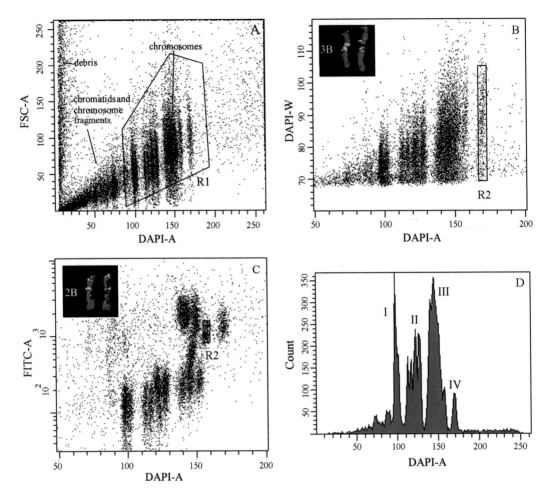

**Fig. 1** Flow cytometric analysis of chromosomes of bread wheat cv. Chinese Spring. (**a**) Dot plot FSC vs. DAPI is used for separating populations of chromosomes (region R1) from chromatids and cellular debris (*arrows*). This gate is applied to the remaining dot plots and histograms. (**b**) Dot plot DAPI-W vs. DAPI-A created for analysis and sorting chromosomes stained with DAPI only. DAPI-A (pulse area) parameter represents total fluorescence of the particle, while DAPI-W (pulse width) parameter corresponds to the length of the particle and is used for discrimination between singlet and doublet events. Region R2 is made for sorting chromosome of interest. *Inset*: images of chromosome 3B sorted from region R2. The chromosome was identified after FISH with probes for GAA microsatellites (*green*) and *Afa* family repeats (*red*). A region of the dot plot was zoomed in to see the populations of chromosomes in more detail. (**c**) Dot plot DAPI-W vs. FITC-A serves to analyze and sort chromosomes simultaneously stained with DAPI and labeled by FISHIS. In this example, wheat chromosomes were labeled with GAA-FITC probe. Region R2 is made for sorting chromosome of interest. *Inset*: images of chromosome 2B sorted from region R2. The chromosome was identified after FISH with probes for GAA microsatellites (*green*) and *Afa* family repeats (*red*). A region of the dot plot was zoomed in to see the populations of chromosomes in more detail. (**d**) Histogram of relative fluorescence intensity of DAPI-stained chromosomes (flow karyotype). It consists of three groups of chromosomes (I-III) and a peak IV representing single-chromosome type (chromosome 3B). Peak I comprises three chromosomes (1D, 4D, 6D), peak II six chromosomes (1A, 6A, 2D, 3D, 5D, 7D), and peak III 11 chromosomes (2A, 3A, 4A, 5A, 7A, 1B, 2B, 4B, 5B, 6B, 7B), respectively

**Table 4**
**Number of sorted chromosomes and their collection for various applications**

| Application | No. of chromosomes | Collection format | Collection liquid |
|---|---|---|---|
| Single-chromosome amplification | 1 | 0.2 mL PCR tube | Reaction mix (3 µL) |
| Physical mapping using PCR | 500–1000 | 0.5 mL PCR tube | RNase-free water (10 µL) |
| Physical mapping using FISH | 1000–2000 | Microscopic slide | P5 buffer (5 µL) |
| NGS sequencing of MDA-amplified chromosomes | 25,000–100,000[a] | 0.5 mL PCR tube | RNase-free water (40 µL) |
| BAC libraries | 200,000[a] | 1.5 mL tube | HKS buffer (220 µL) |
| Optical mapping | 600,000[a] | 1.5 mL tube | HKS buffer (660 µL) |
| Proteomic analysis | 5,000,000 | 15 mL conical tube | LB01 P buffer (1 mL) supplemented with 150 µL PMSF stock solution |

[a]These numbers represent one batch. Total number of batches depends on size of particular chromosome

FITC vs. DAPI dot plot (Fig. 1c) for simultaneous analysis of DNA content and specific DNA sequence labeling by FISHIS. Create histogram for DAPI showing distribution of relative DNA content among chromosomes (Fig. 1d).

5. Run the sample and adjust instrument settings for each parameter so that the populations corresponding to chromosomes are in the field. Analyze at least 20,000 chromosomes and save the data.

6. Create sorting region (R2) surrounding the population of chromosome(s) of interest (Fig. 1b, c). In case of sorting more populations simultaneously, create appropriate sorting regions.

7. Sort required number of chromosomes into appropriate collection vessels or onto microscopic slides depending on subsequent application (Table 4).

***3.7 Estimation of Purity in Sorted Fractions Using FISH (See Note 11)***

1. Sort approximately 2000 chromosomes into 5 µL drop of P5 buffer on a microscope slide. Leave the drop to air dry and keep in the dark at room temperature until use (*see* **Note 12**).

2. Add 25 µL of hybridization mix, place a coverslip and seal with rubber cement.

3. Place the slide at 80 °C for 45 s, letting the chromosomes denature.

4. Move the slide into a humidity chamber and incubate overnight at 37 °C.

5. Transfer the slide into a container filled with 2× SSC and carefully remove the coverslip using tweezers. Wash for 10 min at 42 °C.

6. Wash in 0.1× SSC for 5 min at 42 °C.

7. Incubate in 2× SSC for 10 min at 42 °C.

8. Remove the container from incubator. Replace the solution with heated (42 °C) 2× SSC solution, and incubate the slide for 10 min at room temperature.

9. Wash in 4× SSC for 10 min at room temperature.

10. Remove the slide from the container and put 60 µL of 1% blocking solution over the area with the chromosomes. Cover the slide with a parafilm coverslip and incubate for 10 min at room temperature. Repeat the incubation in blocking solution two more times.

11. Add the solution of fluorescently labeled antibody (follow manufacturer's instructions regarding concentration) in 60 µL of 1% blocking solution, and incubate for 1 h at 37 °C. This step is omitted in the case of directly labeled fluorescent probes.

12. Wash the slide three times in 4× SSC solution for 10 min at 40 °C.

13. Add Vectashield solution containing DAPI and cover with a cover slip.

14. Analyze prepared slide using a fluorescence microscope. For purity check of sorted chromosomes, evaluate at least 100 chromosomes from three independent slides of the same-sorted chromosome.

## 4    Notes

1. Always use viable and healthy seeds. It is advisable to check the germination of every new seed batch before the experiment. The number of seeds necessary for preparation of 1 mL sample depends on the number of roots per seedling and size of root tip.

2. Although the optimal concentration of HU for synchronization of the bread wheat cell cycle is 2.0 mM, in the case of ditelosomic lines, the concentration of 1.25 mM gives better results.

3. In order to achieve good chromosome yields, metaphase index should not be lower than 50%. Metaphase index is determined after microscopic observation of squash preparations of synchronized root tips (not described in this protocol).

4. Overnight treatment in ice-cold water helps in reducing the frequency of chromosome clumps in some species.

5. During chromosome isolation and subsequent analysis and sorting using flow cytometer, the chromosomes are exposed to shearing forces. Mild fixation of roots with formaldehyde makes the chromosomes more resilient. Formaldehyde-fixed roots should be stored on ice and processed within few hours. As formaldehyde is harmful, always wear protecting laboratory gloves and work in biosafety hood.

6. For some applications of sorted chromosomes (preparation of BAC libraries and optical mapping), HKS buffer is used for chromosome isolation instead. In case of proteomic analyses, LB01-P buffer is used.

7. For most plant species, mechanical homogenization is used. However, in species with bigger root tips (e.g., faba bean), higher chromosome yields are obtained when chopping using sharp razor blade.

8. It is advisable to double the amount of root tips needed due to dilution of the chromosome suspension during subsequent treatments.

9. The volume of Tris–HCl solution needed to lower pH of suspension is typically 140–150 µL.

10. Syringing the chromosome suspension just before flow cytometric analysis will decrease the number of chromosome clumps in some species (e.g., faba bean).

11. Estimation of purity in sorted chromosome population is important in all applications. Using FISH, it is possible to estimate not only the percentage of contaminants but also their nature.

12. It is advised to process the slides within few days after sorting as the sucrose present in the buffer could cause dampness of specimen.

## Acknowledgment

This work was supported by grant LO1204 from the National Program of Sustainability I. We also thank Ms. Zdeňka Dubská and Ms. Romana Šperková for excellent technical assistance.

## References

1. Doležel J, Vrána J, Cápal P, Kubaláková M, Burešová V, Šimková H (2014) Advances in plant chromosome genomics. Biotechnol Adv 32:122–136

2. Gray JW, Carrano AV, Steinmetz LL, Van Dilla MA, Moore DH, Mayall BH, Mendelsohn ML (1975) Chromosome measurement and sorting by flow systems. Proc Natl Acad Sci U S A 72:1231–1234

3. Carrano AV, Gray JW, Langlois RG, Burkhart-Schultz KJ, Van Dilla MA (1979) Measurement and purification of human chromosomes by flow cytometry and sorting. Proc Natl Acad Sci U S A 76:1382–1384

4. de Laat AMM, Blaas J (1984) Flow-cytometric characterization and sorting of plant chromosomes. Theor Appl Genet 67:463–467

5. Doležel J, Vrána J, Šafář J, Bartoš J, Kubaláková M, Šimková H (2012) Chromosomes in the flow to simplify genome analysis. Funct Integr Genomics 12:397–416

6. Doležel J, Číhalíková J, Lucretti S (1992) A high-yield procedure for isolation of metaphase chromosomes from root tips of Vicia faba L. Planta 188:93–98

7. Gualberti G, Doležel J, Macas J, Lucretti S (1996) Preparation of pea (Pisum sativum L.) chromosome and nucleus suspensions from single root tips. Theor Appl Genet 92:744–751

8. Kaeppler HF, Kaeppler SM, Lee JH, Arumuganathan K (1997) Synchronization of cell division in root tips of seven major cereal species for high yields of metaphase chromosomes for flow-cytometric analysis and sorting. Plant Mol Biol Rep 15:141–147

9. Lysák MA, Číhalíková J, Kubaláková M, Šimková H, Künzel G, Doležel J (1999) Flow karyotyping and sorting of mitotic chromosomes of barley (Hordeum vulgare L.). Chromosome Res 7:431–444

10. Vrána J, Kubaláková M, Šimková H, Číhalíková J, Lysák MA, Doležel J (2000) Flow-sorting of mitotic chromosomes in common wheat (Triticum aestivum L.). Genetics 156:2033–2041

11. Kubaláková M, Valárik M, Bartoš J, Vrána J, Číhalíková J, Molnár-Láng M, Doležel J (2003) Analysis and sorting of rye (Secale cereale L.) chromosomes using flow cytometry. Genome 46:893–905

12. Lee JH, Arumuganathan K, Yen Y, Kaeppler S, Kaeppler H, Baenziger PS (1997) Root tip cell cycle synchronization and metaphase-chromosome isolation suitable for flow sorting in common wheat (Triticum aestivum L.). Genome 40:633–638

13. Lucretti S, Doležel J (1997) Bivariate flow karyotyping in broad bean (Vicia faba). Cytometry 28:236–242

14. Khlestkina EK (2014) Current applications of wheat and wheat-alien precise genetic stocks. Mol Breeding 34:273–281

15. Giorgi D, Farina A, Grosso V, Gennaro A, Ceoloni C, Lucretti S (2013) FISHIS: fluorescence in situ hybridization in suspension and chromosome flow sorting made easy. PLoS One 8:e57994

16. Lucretti S, Doležel J, Schubert I, Fuchs J (1993) Flow karyotyping and sorting of Vicia faba chromosomes. Theor Appl Genet 85:665–672

17. Neumann P, Lysák M, Doležel J, Macas J (1998) Isolation of chromosomes from Pisum sativum L. hairy root cultures and their analysis by flow cytometry. Plant Sci 137:205–215

18. Gill KS, Arumuganathan K, Lee JH (1999) Isolating individual wheat (Triticum aestivum) chromosome arm by flow cytometric analysis of ditelosomic lines. Theor Appl Genet 98:1248–1252

19. Macas J, Doležel J, Lucretti S, Pich U, Meister A, Fuchs J, Schubert I (1993) Localization of seed protein genes on flow-sorted field bean chromosomes. Chromosome Res 1:107–115

20. Neumann P, Požárková D, Vrána J, Doležel J, Macas J (2002) Chromosome sorting and PCR-based physical mapping in pea (Pisum sativum L.). Chromosome Res 10:63–71

21. Berkman PJ, Skarshewski A, Manoli S, Lorenc MT, Stiller J, Smits L, Lai K, Campbell E, Kubaláková M, Šimková H, Batley J, Doležel J, Hernandez P, Edwards D (2012) Sequencing wheat chromosome arm 7BS delimits the 7BS/4AL translocation and reveals homoeologous gene conservation. Theor Appl Genet 124:423–432

22. Tiwari VK, Wang S, Sehgal S, Vrána J, Friebe B, Kubaláková M, Chhuneja P, Doležel J, Akhunov E, Kalia B, Sabir J, Gill BS (2014) SNP Discovery for mapping alien introgressions in wheat. BMC Genomics 15:273

23. Feuillet C, Eversole K (2007) Physical mapping of the wheat genome: a coordinated effort to lay the foundation for genome sequencing and develop tools for breeders. Isr J Plant Sci 55:307–313

24. Choulet F, Alberti A, Theil S, Glover N, Barbe V, Daron J, Pingault L, Sourdille P, Couloux A, Paux E, Leroy P, Mangenot S, Guilhot N, Le Gouis J, Balfourier F, Alaux M, Jamilloux V, Poulain J, Durand C, Bellec A, Gaspin C, Šafář J, Doležel J, Rogers J, Vandepoele K, Aury JM, Mayer K, Berges H, Quesneville H, Wincker P, Feuillet C (2014) Structural and functional partitioning of bread wheat chromosome 3B. Science 345:1249721

25. The International Wheat Genome Sequencing Consortium (IWGSC) (2014) A chromosome-based draft sequence of the hexaploid bread wheat (Triticum aestivum) genome. Science 345:1251788

26. Valárik M, Bartoš J, Kovářová P, Kubaláková M, de Jong H, Doležel J (2004) High-resolution FISH on super-stretched flow-sorted plant chromosomes. Plant J 37:940–950

27. Kopecký D, Martis M, Číhalíková J, Hřibová E, Vrána J, Bartoš J, Kopecká J, Cattonaro F, Stočes Š, Novák P, Neumann P, Macas J, Šimková H, Studer B, Asp T, Baird JH, Navrátil P, Karafiátová M, Kubaláková M, Šafář J, Mayer K, Doležel J (2013) Flow sorting and sequencing meadow fescue chromosome 4F. Plant Physiol 163:1323–1337

28. Kejnovský E, Vrána J, Matsunaga S, Souček P, Široký J, Doležel J, Vyskot B (2001) Localization of male-specifically expressed *MROS* genes on *Silene latifolia* by PCR on flow-sorted sex chromosomes and autosomes. Genetics 158:1269–1277

29. Vláčilová K, Ohri D, Vrána J, Číhalíková J, Kubaláková M, Kahl G, Doležel J (2002) Development of flow cytogenetics and physical genome mapping in chickpea (*Cicer arietinum* L.). Chromosome Res 10:695–706

30. Šimková H, Svensson JT, Condamine P, Hřibová E, Suchánková P, Bhat PR, Bartoš J, Šafář J, Close TJ, Doležel J (2008) Coupling amplified DNA from flow-sorted chromosomes to high-density SNP mapping in barley. BMC Genomics 9:294

31. Vrána J, Šimková H, Kubaláková M, Číhalíková J, Doležel J (2012) Flow cytometric chromosome sorting in plants: the next generation. Methods 57:331–337

32. Hernandez P, Martis M, Dorado G, Pfeifer M, Gálvez S, Schaaf S, Jouve N, Šimková H, Valárik M, Doležel J, Mayer KFX (2012) Next-generation sequencing and syntenic integration of flow-sorted arms of wheat chromosome 4A exposes the chromosome structure and gene content. Plant J 69:377–386

33. Kantar M, Akpınar BA, Valárik M, Lucas SJ, Doležel J, Hernández P, Budak H (2012) Subgenomic analysis of microRNAs in polyploid wheat. Funct Integr Genomics 12:465–479

34. Shatalina M, Wicker T, Buchmann JP, Oberhaensli S, Šimková H, Doležel J, Keller B (2013) Genotype-specific SNP map based on whole chromosome 3B sequence information from wheat cultivars Arina and Forno. Plant Biotechnol J 11:23–32

35. Martis MM, Klemme S, Banaei-Moghaddam AM, Blattner FR, Macas J, Schmutzer T, Scholz U, Gundlach H, Wicker T, Šimková H, Novák P, Neumann P, Kubaláková M, Bauer E, Haseneyer G, Fuchs J, Doležel J, Stein N, Mayer KFX, Houben A (2012) Selfish supernumerary chromosome reveals its origin as a mosaic of host genome and organellar sequences. Proc Natl Acad Sci U S A 109:13343–13346

36. Martis MM, Zhou R, Haseneyer G, Schmutzer T, Vrána J, Kubaláková M, König S, Kugler KG, Scholz U, Hackauf B, Korzun V, Schön CC, Doležel J, Bauer E, Mayer KFX, Stein N (2013) Reticulate evolution of the rye genome. Plant Cell 25:3685–3698

37. Šimková H, Číhalíková J, Vrána J, Lysák MA, Doležel J (2003) Preparation of HMW DNA from plant nuclei and chromosomes isolated from root tips. Biol Plant 46:369–373

38. Zatloukalová P, Hřibová E, Kubaláková M, Suchánková P, Šimková H, Adoración C, Kahl G, Millán T, Doležel J (2011) Integration of genetic and physical maps of the chickpea (*Cicer arietinum* L.) genome using flow-sorted chromosomes. Chromosome Res 19:729–739

39. Kofler R, Bartoš J, Gong L, Stift G, Suchánková P, Šimková H, Berenyi M, Burg K, Doležel J, Lelley T (2008) Development of microsatellite markers specific for the short arm of rye (*Secale cereale* L.) chromosome 1. Theor Appl Genet 117:915–926

40. Wenzl P, Suchánková P, Carling J, Šimková H, Huttner E, Kubaláková M, Sourdille P, Paul E, Feuillet C, Kilian A, Doležel J (2010) Isolated chromosomes as a new and efficient source of DArT markers for the saturation of genetic maps. Theor Appl Genet 121:465–474

41. Šimková H, Šafář J, Kubaláková M, Suchánková P, Číhalíková J, Robert-Quatre H, Azhaguvel P, Weng Y, Peng J, Lapitan NLV, Ma Y, You FM, Luo M-C, Bartoš J, Doležel J (2011) BAC libraries from wheat chromosome 7D: efficient tool for positional cloning of aphid resistance genes. J Biomed Biotechnol 2011: 302543

42. Šafář J, Bartoš J, Janda J, Bellec A, Kubaláková M, Valárik M, Pateyron S, Weiserová J, Tušková R, Číhalíková J, Vrána J, Šimková H, Faivre-Rampant P, Sourdille P, Caboche M, Bernard M, Doležel J, Chalhoub B (2004) Dissecting large and complex genomes: flow sorting and BAC cloning of individual chromosomes from bread wheat. Plant J 39:960–968

43. Janda J, Šafář J, Kubaláková M, Bartoš J, Kovářová P, Suchánková P, Pateyron S, Číhalíková J, Sourdille P, Šimková H, Faivre-Rampant P, Hřibová E, Bernard M, Lukaszewski A, Doležel J, Chalhoub B (2006) Advanced resources for plant genomics: BAC library specific for the short arm of wheat chromosome 1B. Plant J 47:977–986

44. Šimková H, Šafář J, Suchánková P, Kovářová P, Bartoš J, Kubaláková M, Janda J, Číhalíková J, Mago R, Lelley T, Doležel J (2008) A novel resource for genomics of Triticeae: BAC library specific for the short arm of rye (*Secale cereale* L.) chromosome 1R (1RS). BMC Genomics 9:237

45. Petrovská B, Jeřábková H, Chamrád I, Vrána J, Lenobel R, Uřinovská J, Šebela M, Doležel J (2014) Proteomic analysis of barley cell nuclei

purified by flow sorting. Cytogenet Genome Res 143:78–86

46. Molnár I, Kubaláková M, Šimková H, Cseh A, Molnár-Láng M, Doležel J (2011) Chromosome isolation by flow sorting in Aegilops umbellulata and Ae comosa and their allotetraploid hybrids Ae biuncialis and Ae geniculata. PLoS One 6:e27708

47. Králová T, Čegan R, Kubát Z, Vrána J, Vyskot B, Vogel I, Kejnovský E, Hobza R (2014) Identification of a novel retrotransposon with sex chromosome-specific distribution in *Silene latifolia*. Cytogenet Genome Res 143:87–95

48. Kubaláková M, Kovářová P, Suchánková P, Číhalíková J, Bartoš J, Lucretti S, Watanabe N, Kianian SF, Doležel J (2005) Chromosome sorting in tetraploid wheat and its potential for genome analysis. Genetics 170:823–829

49. Kovářová P, Navrátilová A, Macas J, Doležel J (2007) Chromosome analysis and sorting in *Vicia sativa* using flow cytometry. Biol Plant 51:43–48

# Chapter 11

## Construction of BAC Libraries from Flow-Sorted Chromosomes

Jan Šafář, Hana Šimková, and Jaroslav Doležel

### Abstract

Cloned DNA libraries in bacterial artificial chromosome (BAC) are the most widely used form of large-insert DNA libraries. BAC libraries are typically represented by ordered clones derived from genomic DNA of a particular organism. In the case of large eukaryotic genomes, whole-genome libraries consist of a hundred thousand to a million clones, which make their handling and screening a daunting task. The labor and cost of working with whole-genome libraries can be greatly reduced by constructing a library derived from a smaller part of the genome. Here we describe construction of BAC libraries from mitotic chromosomes purified by flow cytometric sorting. Chromosome-specific BAC libraries facilitate positional gene cloning, physical mapping, and sequencing in complex plant genomes.

**Key words** BAC library, BAC vector, Chromosomes, DNA cloning, High molecular weight (HMW) DNA, Pulse-field gel electrophoresis (PFGE), Clone

## 1 Introduction

Bacterial artificial chromosome (BAC) libraries are indispensable resources for plant genetic and genomic studies. They are successfully employed in positional gene cloning, construction of BAC contig physical maps, DNA marker development, and genome sequencing. BAC-derived markers and probes enable linking physically defined genomic regions with genetic mapping of loci underlying traits of interest. This feature also facilitates map-based cloning of genes responsible for specific phenotypes. BAC clones are also used in cytogenetics as probes for fluorescence in situ hybridization, in comparative and evolutionary studies, as well as in physical mapping [1].

BAC libraries remain invaluable in genome-sequencing projects. Although the recent expansion of next-generation sequencing (NGS) platforms stimulated whole-genome shotgun sequencing approaches, sequence assembling in complex genomes with preponderance of repetitive DNA and polyploid nature still

Shahryar F. Kianian and Penny M.A. Kianian (eds.), *Plant Cytogenetics: Methods and Protocols*,
Methods in Molecular Biology, vol. 1429, DOI 10.1007/978-1-4939-3622-9_11,
© Springer Science+Business Media New York 2016

remains a challenge [2]. The utility of BAC libraries for genome sequencing was proven in several important projects [3–5]. Recent improvements in the clone-by-clone sequencing strategy, which include sequencing single BAC clones or BAC pools by NGS, provide a realistic chance to tackle huge and complex plant genomes and generate high-quality gold-standard reference sequences [6]. BAC libraries have been constructed from many plant species in a number of research laboratories and by specialized commercial companies. Even though protocols for BAC library construction are published [7, 8], the procedure of creating BAC libraries is not trivial in plants. Their construction is mainly hampered by difficulties in preparing quality high-molecular weight (HMW) DNA—a task particularly challenging in plants. This is due to specific features of plant cells such as rigid cell wall, the presence of carbohydrates, and various secondary metabolites, which may bind to DNA decreasing its accessibility by restricting enzymes and compromising cloning ability. Traditionally, BAC libraries were constructed from genomic DNA prepared from nuclei released from tissues by mechanical homogenization in liquid nitrogen and purified by gradient centrifugation [9]. We have pioneered an alternative approach in which the libraries are constructed from nuclei or chromosomes isolated by mechanical homogenization of tissues fixed mildly by formaldehyde and purified by flow cytometric sorting [10]. This approach results in high-quality HMW DNA almost free of organelle DNA and secondary metabolites [11]—a prerequisite for construction of high-quality BAC libraries. This protocol overcomes problems caused by high levels of polyphenols and polysaccharides present in plant species such as banana [11] and olive tree [12]. An important advantage of using flow cytometry is the possibility of purifying individual chromosome types and cloning them one by one into the BAC vector. In comparison with genomic BAC libraries, chromosome libraries comprise smaller and hence manageable number of clones. For instance, a whole-genome BAC library of bread wheat representing 9.3× genome coverage consists of more than 1,200,000 clones [13], while the number of clones for one chromosome-/chromosome arm-specific library with 15× coverage ranges from 50 to 90 thousand [14]. Moreover, slicing a genome into smaller well-defined parts enables splitting the work among several collaborating teams. Thus, generation of a ready-to-sequence physical map and sequencing of a complex genome is accelerated.

To date, a complete set of chromosome and chromosome arm-specific BAC libraries (45 in total) were constructed for cultivar Chinese Spring of bread wheat [15]. This includes a 3B- and 1BS-specific library to facilitate positional cloning [16, 17]. Additional BAC libraries for the rye chromosome 1S [18], wild wheat relative *Aegilops umbellulata* chromosome 6U, and *Pisum sativum* chromosome 5 have been produced (unpublished results).

The protocols in this chapter describe the construction of chromosome-specific BAC libraries from flow-sorted chromosomes and chromosomal arms. Briefly, HMW DNA is prepared from flow-sorted chromosomes; their DNA is partially digested and size selected by pulse-field gel electrophoresis. Afterwards, DNA fragments are released from the gel by electroelution and ligated into a BAC vector. The ligation mixture is desalted and used to transform electrocompetent *Escherichia coli* cells. The bacterial clones with recombinant molecules are picked and stored in 384-well plates.

# 2   Materials

Prepare all solutions using ultrapure water and analytical grade reagents. Prepare and store all reagents at room temperature (unless indicated otherwise). Diligently follow all waste disposal regulations when disposing waste materials.

## 2.1   Solutions

1. Isolation buffer (IB): 10-mM Tris–HCl, 10-mM Na$_2$EDTA, 1-mM spermine, 1-mM spermidine 130-mM KCl 20-mM NaCl, 1% Triton X-100, and pH 9.4. Prepare fresh from stock solution (*see* **Note 1**).

2. 2% low-melting agarose in 1× IB.

3. Lysis buffer B: 0.5-M EDTA, 1% w/v sodium lauryl sarcosine, and pH 8.0.

4. Lysis buffer C: 0.5-M EDTA, 1% w/v sodium lauryl sarcosine, and pH 9.0.

5. ET buffer: 1-mM Tris base, 50-mM EDTA, and pH 8.0.

6. TE buffer: 10-mM Tris–HCl, 1-mM EDTA, and pH 8.0.

7. 0.25× TBE buffer: 22.5-mM Tris base, 22.5-mM boric acid, 0.5-mM EDTA, and pH 8.0.

8. 1.3× TAE buffer: 0.52-M Tris base, 0.52-M acetic acid, 1.3-mM EDTA, and pH 8.0.

9. Digestion buffer: 50-mM potassium acetate, 20-mM Tris acetate, 10-mM magnesium acetate, 2-mM DTT, 4-mM spermidine, 0.39-mg/mL BSA, and pH 7.9.

10. Desalting gel: 1% agarose gel and 20-mM glucose.

11. 2YT medium: 1.6% tryptone (16.0 g/L), 1.0% yeast extract (10.0 g/L), 0.5% NaCl (5.0 g/L), and pH 7.0.

12. Solid 2YT medium: 1.6% tryptone (16.0 g/L), 1.0% yeast extract (10.0 g/L), 0.5% NaCl (5.0 g/L), 1.6% agar (16.0 g/L), and 0.025% v/v chloramphenicol stock solution (*see* **Note 2**).

13. Selection 2YT medium: 2YT medium, 1.6 % w/v agar, 0.45 % v/v X-GAL stock solution, 0.045 % (v/v) IPTG stock solution, and 0.025 % v/v chloramphenicol stock solution.

14. Freezing medium: 2YT medium and 6 % glycerol.

15. Chloramphenicol stock solution: 5 % w/v dissolved in 96 % ethanol. Store the stock solution at –20 °C for several weeks.

16. X-GAL     (5-bromo-4-chloro-3-indolyl-beta-d-galactoside) stock     solution:     2 %     w/v     X-GAL     dissolved     in NN-dimethylformamide.

17. IPTG (isopropylthiogalactoside) stock solution: 20 % w/v in water.

18. 2-M glucose stock solution.

19. Ethidium bromide solution: 1 % w/v in water (*see* **Note 3**).

20. Proteinase K: Concentration of 1-mg/mL in water.

21. Blue juice: 0.05 % w/v bromophenol blue, 42.5 % v/v glycerol, and 100 mM EDTA (*see* **Note 4**).

## 2.2  Special Reagents

1. Cloning-ready BAC vector (*see* **Note 5**).

2. Restriction endonuclease *Hin*dIII with 10× buffer and 100× BSA (*see* **Note 6**).

3. T4 ligase with 10× ligase buffer.

4. Competent cells: *Escherichia coli* MegaX DH10B T1 competent cells (*see* **Note 7**).

5. Recovery medium (*see* **Note 8**).

6. Lambda phage DNA (set of concentrations ranging from 1 ng/μL to 10 ng/μL).

7. Pulse-field gel electrophoresis (PFGE) lambda ladder.

## 2.3  Supplies

1. Miracloth (50-μm mesh size).

2. Polystyrene cuvettes with conical bottom and cap.

3. Agarose plug molds (e.g., from BioRad).

4. Green membrane caps (e.g., from BioRad).

5. 384-well microtiter plates (*see* **Note 9**).

6. Bacterial plating trays (e.g., Q-trays from Genetix).

7. Plating rod (*see* **Note 10**).

## 2.4  Equipment

1. Pulse-field gel electrophoresis (PFGE): (e.g., the BioRad CHEF-DR II system, the CHEF-DR III system, and the CHEF Mapper™ XA system), gel casting stand, comb, and accessories.

2. Electroporation system with booster (*see* **Note 11**).

3. Automated microtiter plate filler (e.g., Q-fill2 or Q-fill3 from Genetix).

4. Electroelution system with membrane caps (e.g., BioRad Model 422 Electro-Eluter, *see* **Note 12**).

5. A robotic workstation for bacteria handling (colony picking, replicating, re-arraying).

## 3  Methods

Carry out all procedures at 4 °C unless specified otherwise.

### 3.1  Preparation of HMW DNA from Flow-Sorted Chromosomes

1. Sort 200,000 chromosomes into a 1.5-mL polystyrene cuvette filled with 320 μL of 1.5× isolation buffer (IB) (*see* **Note 13**).

2. Pellet the chromosomes by centrifugation in a swinging-bucket rotor at $300 \times g$ and 4 °C for 30 min with slow braking at completion of centrifugation.

3. Carefully remove by gentle pipetting the entire volume of supernatant except for 12 μL. Vortex gently to resuspend the chromosomes in the remaining IB solution in the bottom of the cuvette, and prewarm to 50 °C in a water bath for 5–10 min.

4. Add 8 μL of melted 2 % low-gelling agarose in 1× IB that is preheated to 50 °C. Vortex gently and then incubate in a water bath for 5 min (*see* **Note 14**).

5. Pipette the chromosomes embedded in agarose with a cutoff pipette tip into a prechilled plug mold. Let solution solidify at 4 °C for 5–10 min to form a miniplug (*see* **Note 15**).

6. Incubate the miniplugs for 24 h in lysis buffer C using 0.5 mL/plug and supplement with freshly prepared proteinase K at a final concentration of 0.1 mg/mL in lysis buffer. Incubate at 37 °C in orbital shaker adjusted to 50 rpm (*see* **Note 16**).

7. Exchange for the same amount of lysis buffer B plus proteinase K, and incubate for another 24 h at 37 °C in orbital shaker.

8. Wash the miniplugs with several mL of ET buffer and store in ET buffer at 4 °C for as long as half a year (*see* **Note 17**).

### 3.2  Partial Digestion of HMW DNA

1. Prior to the partial digestion of chromosomal DNA, the agarose miniplugs must be washed six to eight times, for 1 h each, in 8 mL of ice-cold TE buffer.

2. Cut miniplugs into three slices each with a glass microscope coverslip, and transfer them into 1.5 mL microcentrifuge tubes placing three slices per tube.

3. Add 200 μL of ice-cold digestion buffer (DB) into each tube.

4. Incubate for 1 h on ice under gentle shaking.

5. Remove the DB and add 97 μL of fresh DB into each tube. Add various amounts of diluted restriction endonuclease (e.g., *Hind*III) (*see* **Note 18**).

6. Incubate tubes on ice for 1 h under gentle shaking.

7. Fix the tubes in a foam float and transfer into a water bath preheated to 37 °C. Incubate for exactly 20 min (*see* **Note 18**).

8. Stop the digestion by transferring the tube to ice and adding 30 μL 0.5-M EDTA (pH 8.0).

9. Collect the slices of miniplugs from all tubes in one Petri dish and add several mL of ice-cold 0.25× TBE. Wash the slices three times, for 10 min each, in 0.25× TBE on ice under gentle shaking.

**3.3 Size Selection of the Digested DNA Fragments**

*First size selection*

1. Prepare a 1 % agarose gel in 0.25× TBE.

2. After solidification, create an extended well by excising boundaries between three wells of the gel.

3. Load all miniplug slices into the extended well. Load two slices of the PFGE lambda ladder size standard into wells flanking the extended well, leaving one empty well between the extended well and the ladder wells. There should be four wells with ladder, two wells on the left and two on the right of the gel (Fig. 1). Fill up all loaded wells with 1 % agarose in 0.25× TBE.

4. Perform pulse-field gel electrophoresis (PFGE) in prechilled 0.25× TBE under the following conditions: temperature 12.5 °C, voltage 6 V/cm, switch time 1–50 s, and run time 18 h (*see* **Note 19**).

5. Excise the central part (Fig. 1) of the gel containing the digested DNA, pour several milliliters of 0.25× TBE buffer over the gel, and store in a fridge.

6. Stain the flanking parts of the gel containing the size marker in 0.5-μg/mL ethidium bromide for 30 min (*see* **Note 20**). Then photograph the gel with a ruler.

7. Delimit on the non-stained part of the gel a zone ranging from 100 to 300 kb using the size marker and the ruler on the photograph as references and excise and divide the gel zone horizontally into three equal sections (bottom (B), middle(M), top(T)) (*see* **Note 21**).

*Second size selection*

1. Prepare a new 0.9 % agarose gel in 0.25× TBE.

2. After solidification, create one big sample well of suitable size for gel pieces containing the B, M, and T fraction.

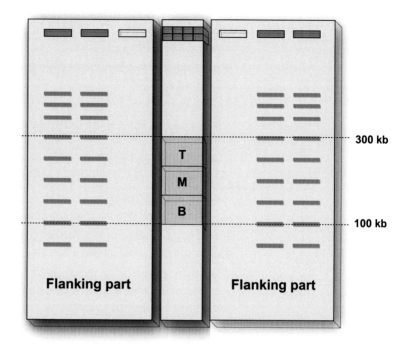

**Fig. 1** Agarose gel at first size selection. The DNA fragments are size selected after separation by pulse-field gel electrophoresis. The gel is sliced into three pieces; flanking parts are stained by ethidium bromide to visualize lambda ladder and photographed with an attached ruler. A zone ranging from 100 to 300 kb in the central, non-stained part of the gel is delimited based on estimated positions of corresponding ladder bands. The selected gel zone is excised and horizontally divided into three equal sections (*B* bottom, *M* middle, *T* top)

3. Load the pieces of gel containing the B, M, and T fraction, respectively, into the sample well and two slices of the size standard into wells flanking the sample well on both sides. Subsequently, load used wells with some agarose and let solidify.

4. Perform pulse-field gel electrophoresis in prechilled 0.25× TBE under the following conditions: temperature 12.5 °C, voltage 6 V/cm, switch time 3 s, and run time 18.0 h.

5. Excise part of the gel containing the B and M fraction and put them into a fridge suffused by several mL of 1.3× TAE buffer.

6. Stain the flanking parts of the gel containing the size marker and the T fraction in 0.5-µg/mL ethidium bromide. Photograph with a ruler.

7. Locate the zone from 100 to 150 kb for the B fraction and 150–225 kb for the M fraction. Excise these regions. Divide the B-containing zone horizontally into two equal sections (B1 and B2) (*see* **Note 22**).

**3.4 Electroelution**

1. Put the gel slices with the DNA to be eluted into a 50-mL tube containing 30 mL 1.3× TAE buffer. Wash for half an hour under gentle shaking on ice.

2. Exchange the buffer and repeat the washing for three additional times.

3. In the meantime, excise several circles (2-cm in diameter) of Miracloth and soak them in 1.3× TAE buffer at 65 °C for 1 h.

4. Take the required number of membrane caps for the electroelution unit, and soak them in 1.3× TAE buffer at 65 °C for 1 h.

5. Assemble electroelution cuvette(s) of the Electro-Eluter (Fig. 2). Fill the cuvettes up with chilled 1.3× TAE buffer.

6. Insert each cuvette into a rubber gasket on the tube rack. Use gasket plugs to seal gaskets that do not contain cuvettes.

7. Insert the gel piece containing the size-selected DNA into the cuvette, and push it gently by using the wide side of a pipette tip to the bottom of the cuvette.

8. Fill the upper and lower tank with chilled 1.3× TAE, avoiding the formation of bubbles on the membrane cap.

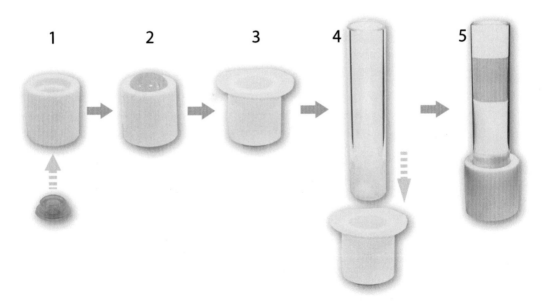

**Fig. 2** Assembling of the electroelution device. A membrane cap is joined with a silicone adaptor (*1*). The joint assembly unit is filled up with the electroelution buffer to form reverse meniscus (*2*). A Miracloth circle is placed on the top of the meniscus (*3*). A glass tube is inserted into the assembly unit (*4*). The tube is filled up with cold elution buffer, and the gel slice with DNA is lowered to the bottom of the tube using the wide end of a pipette tip (*5*). Please, see a manufacturer's instructions for more details and for preparation of electroelution aperture itself

9. Run the electroelution at constant current (10 mA per cuvette) for 70–80 min (*see* **Note 23**).

10. Interchange electrodes, and let the DNA run in opposite direction for 90 s to release DNA molecules stuck to the membrane.

11. Gently remove the entire volume of the buffer above the Miracloth from the cuvette. Then carefully remove the membrane cap from the silicon adaptor.

12. Pipette gently the DNA-containing droplet with a cutoff or large orifice pipette tip from the membrane cap and transfer to a tube. For volumes exceeding 60 µL, split the entire volume into two tubes.

13. Using an agarose gel, estimate the concentration of the eluted DNA by comparing the amount of sample DNA in 5-µL volume to a concentration series of phage lambda from 5 to 30 ng. Use gel estimate to calculate the total amount of DNA obtained from the electroelution.

### 3.5 Ligation

1. Prepare a ligation mixture consisting of size-selected DNA (volume 30–60 µL) and digested and dephosphorylated pIndigoBAC-5 vector corresponding to the complementary enzyme used in **step 5** of Subheading 3.2, 1/10 volume 10× ligase buffer (*see* **Note 24**), 1–1.5 µL T4 DNA ligase.
   The amount of vector is adjusted to comply with the following weight ratios:

   – Fraction B1: 3.5 ng DNA/1 ng vector.
   – Fraction B2: 4 ng DNA/1 ng vector.
   – Fraction M: 6 ng DNA/1 ng vector.

2. Incubate overnight (optimum 12–13 h) at 16 °C (*see* **Note 25**).

### 3.6 Desalting

1. Prepare 1% agarose gel in 20 mL of deionized water, cool briefly, and add 1-mL 1-M glucose before solidification. Pipette about 1.2 mL of the solution into a 1.5-mL tube. Insert another 1.5-mL tube and let solidify.

2. Remove the inserted tube and let dry in a laminar flow hood for 45 min.

3. Gently pipette the ligation mixture with a cutoff tip into the formed agarose pit. Incubate on ice for 70–75 min (*see* **Note 26**).

4. Gently remove the desalted mixture with a cutoff tip from the tube and record the volume.

### 3.7 Transformation

1. Mix the desalted ligation mixture with electrocompetent *E. coli* cells (e.g., MegaX DH10B T1, Invitrogen) at the ratio of 16 µL mixture per 100 µL cells. Incubate on ice for 5–10 min (*see* **Note 27**).

2. Transform the electrocompetent cells with the recombinant vector by electroporation (15 μL of transformation mixture per electroporation cuvette). If using the Cell-Porator and Voltage Booster System (Life Technologies), apply the following settings:

| Cell-Porator | Voltage Booster System |
|---|---|
| Voltage | 330–340 V |
| Resistance | 4 kΩ |
| Capacitance | 330 μF |
| Impedance | Low ohms |
| Charge rate | Fast |

3. Collect the transformed cells from the cuvettes (total volume typically 340–460 μL) into two bacterial cultivation tubes containing 2–2.5-mL recovery medium each.

4. Incubate at 37 °C under shaking at 175 rpm for 1 h.

5. Plate aliquots using a plating rod to spread the cultivated cells on 22 × 22-cm bacterial-plating trays containing selection 2YT medium to determine the titer of the transformation reaction (*see* **Note 28**). The volume of cultivated cells used per plate varies between 50 and 200 μL depending on DNA size fraction and is supplemented by recovery medium to reach plating volume of 600 μL.

6. Dry for 1 min in laminar flow hood and then incubate the plates bottom up at 37 °C for about 20 h. Store the remaining volume of transformed bacteria suspension in the fridge overnight.

7. Check the density of colonies growing on the plating trays to determine colony-forming unit (CFU) (*see* **Note 29**), and estimate proportion of the blue and white ones. Determine the optimal titer for massive plating of the stored volume of bacterial suspension (*see* **Note 30**).

8. Take the tubes with the bacterial suspension out of the fridge, and warm them up at 37 °C under shaking at 175 rpm for maximum 45 min.

9. Based on CFU previously observed, combine a volume of cells corresponding to approximately 1500–2000 colonies with appropriate amount of recovery medium to reach plating volume of 600 μL, and spread the mixture on plating trays with 2YT selection medium.

10. Incubate the plates bottom up at 37 °C overnight (*see* **Note 31**).

11. Randomly select at least 100 BAC clones for insert determination (*see* **Note 32**).

12. Pick white bacterial colonies of appropriate shape and size using a robotic workstation (e.g., Q-bot, Genetix), and order them in 384-well plates filled with freezing medium (*see* **Note 33**).

13. Incubate the inoculated 384-well plates at 37 °C for at least 16 h for bacterial growth (*see* **Note 34**).

14. Collect and order the required amount of clones to reach desired BAC library coverage (*see* **Note 35**).

15. Prepare a replica of the master copy of the library (*see* **Note 36**).

16. Store the library at –80 °C.

## 4 Notes

1. Preparation of 50-mL 10× isolation buffer (IB): 606 mg Tris (100 mM), 1.86-g Na$_2$EDTA (100 mM), 174-mg spermine·4HCl (10 mM), 127.3-mg 10-mM spermidine·3HCl (10 mM), 4.84-g KCl (1.3 M), 0.234-g NaCl (200 mM), and 500-μL Triton X-100 (1%). Adjust volume to 50 mL with deionized H$_2$O and pH to 9.4. Store at 4 °C for several months.

2. Let 2YT medium cool down after autoclaving to approximately 40 °C prior to adding the chloramphenicol solution.

3. Ethidium bromide should be handled with extra care, because it is a mutagen/carcinogen.

4. Store aliquots at –20 °C. Dilute before use 1:5 with your sample.

5. Cloning-ready vector is fully digested with the same restriction endonuclease as the DNA to be inserted, i.e., with *Hin*dIII. The 5′ end of the vector is dephosphorylated with phosphatase to prevent self-ligation. Preparation of a highly effective vector is a crucial step in the library construction. A detailed protocol for vector preparation has been described in (*see* ref. 7).

6. The use of alternative restriction endonucleases depends upon the sequence of the polycloning site of the particular BAC vector. Such endonucleases as *Bam*HI, *Eco*RI, and *Sph*I are also applicable. We prefer *Hin*dIII enzyme for its low star activity and consequent high cloning efficiency. A recommended supplier is New England Biolabs.

7. There are several bacterial strains convenient for transformation by BAC vector. We recommend using *Escherichia coli* MegaX DH10B T1 strain, distinguished by high cloning efficiency and resistance to bacteriophages T1 and T5.

8. Recovery medium is provided with the set of MegaX DH10B T1 electrocompetent cells. SOC medium (*see* ref. 19) can be used as an alternative.

9. 384-well plates can be purchased from several suppliers. Make sure the selected type is compatible with your robotic workstation. We recommend using microtiter plates from Genetix for their long-term shape stability at both −80 °C and room temperature.

10. Plating rod is a plastic or glass tool for spreading the bacterial culture on agar surface.

11. There are currently a variety of commercially available electroporation devices. Systems with voltage booster are advised for construction of BAC libraries. The Life Technologies BRL Cell-Porator is specifically recommended for use with DH10B cells resulting in higher transformation efficacy.

12. Alternatively, dialysis tube system or gelase treatment can be used, but quality of the released DNA is usually lower.

13. See chapter by J. Vrána et al. within book for the flow-sorting procedure. The amount of 1.5× IB buffer is adjusted based on number of flow-sorted chromosomes and sorted droplet volume, which is instrument-specific. The final ratio of flow-sorted volume to the volume of buffer in cuvette should be 1:1 to reach final concentration of IB 0.75×. Cumulative amount of DNA (derived from number and DNA content of flow-sorted chromosomes) for a successful BAC library construction should be around 5 μg. The flow-sorting approach can also be used for nuclei isolation and purification, and is beneficial in species with a high content of metabolic compounds.

14. We preferentially use low-gelling InCert Agarose, Lonza.

15. Pipette the mixture of chromosomes and agarose promptly to prevent agarose solidification in the pipette tip. Also, use large orifice or cutoff pipette tip to transfer large molecules of DNA to prevent shearing.

16. One aliquot of freshly prepared proteinase K can be stored at 4 °C for usage with lysis buffer B on the next day.

17. We have not observed degradation of HMW DNA as long as 1 year after miniplug preparation. We usually leave miniplugs in TE buffer for at least 2 weeks prior to partial digestion.

18. The restriction endonuclease amount is optimized in a test digestion that precedes the library construction. For flow-sorted chromosomes, the amount of *Hin*dIII varies between 0.1 and 0.2 U per tube. Digestion for 20 min compared with

shorter incubation time gives more reproducible results. Apply the same reagents and digestion time (20 min) for both the optimization and the library construction.

19. The amount of 0.25× TBE buffer is adjusted according to the thickness of the gel. Run time is instrument dependent and can be prolonged to 19 h.

20. Ethidium bromide-staining solution is prepared by 1000× dilution of the ethidium bromide stock solution.

21. Fraction B ("bottom"), fraction M ("middle"), and fraction T ("top") carry DNA fragments 100–150 kbp, 150–225 kbp, and 225–300 kbp, respectively.

22. Gel slices with digested DNA can be stored at 4 °C for several weeks. We recommend initially cloning only one fraction (B1 or B2). The outcomes of cloning and transformation (cloning efficiency, percentage of empty clones) help fine-tune the ligation conditions, namely, DNA/vector ratio, for the other fractions.

23. 70–80 min of electroelution are adequate for fractions B1 and B2. The time can be prolonged to 90 min for fraction M containing larger fragments. Note that the voltage should not exceed 80 V.

24. DTT in the ligation buffer can precipitate at –20 °C. Therefore, the buffer needs to be warmed up at room temperature and vortexed properly prior to use.

25. We usually incubate the ligation mixture in a PCR cycler. Chilled water bath or other incubators are also convenient.

26. Extension of desalting time leads to reduction of sample volume. ~10 % volume reduction is usual.

27. Incubation of the mixture on ice prior to the transformation slightly increases the transformation efficiency.

28. Keep spreading the bacterial suspension as long as you observe wet spots, and the rod is sliding smoothly on the surface of the medium.

29. Colony-forming unit (CFU) is a rough estimate of the number of viable bacteria cells in a sample. Results are usually reported as CFU/mL (colony-forming units per milliliter).

30. Prepare the appropriate amount of bacterial-plating trays (e.g., Q-trays) with selective 2YT medium. See a visual guide of Peterson et al. (*see* ref. 7), for easier maintenance of many Q-trays at once.

31. The incubation time should be adjusted to obtain a majority of clones with diameter of 1–2 mm. Q-trays with bacteria that will not be picked on the following day should be incubated at

37 °C for several hours and subsequently stored at 4 °C until further use (maximum 3 weeks). Just before picking, these bacteria are incubated at 37 °C for additional 10 h.

32. Use BAC miniprep protocol (*see* ref. 7, 13) for BAC plasmids isolation from bacterial cultures. Digest isolated BACs with *Not*I endonuclease to release inserts from vector. Pulse-field gel electrophoresis (PFGE) is used to check and determine the presence and length of inserts. Empty clones (BACs without inserts) should not represent more than 5 %.

33. 384-well plates are filled with 75 μL/well (three fourth of well capacity) of freezing medium by a liquid-handling machine (e.g., Q-fill3, Genetix). Stock of filled plates can be stored at −20 °C for several weeks prior to use.

34. To prevent evaporation, the 384-well plates should be wrapped in a foil during incubation. Density of cell culture can be evaluated visually. Uniform density and viability of bacteria are important for downstream applications such as replication and preparation of high-density filters and influence the library life span.

35. The number of genome/chromosome equivalents covered by the library (library coverage) is calculated based on number of clones, mean insert size of the clones, and genome/chromosome size [coverage = $N \times L/G$, where $N$ = number of clones; $L$ = mean insert size (in bp.); $G$ = genome/chromosome size (in bp.)]. Coverage of 6× is sufficient for most applications except physical map construction, which requires at least 10× coverage. An ordered BAC library consists of numbered and labeled 384-well plates.

36. Use hand-held colony replicator or robotic workstation for preparing replicas of the library. Preparation of one back-up and one working copy is recommended. Once the replicated copies have been made, the "master copy" is stored and not used unless absolutely necessary such as its copies are damaged or lost.

## Acknowledgment

This work was supported by grant LO1204 from the National Program of Sustainability I. We also thank Dr. Zbyněk Milec for creating figures and critical reading of the manuscript.

# References

1. Shearer LA, Anderson LK, de Jong H, Smit S, Goicoechea JL, Roe BA, Hua A, Giovannoni JJ, Stack SM (2014) Fluorescence in situ hybridization and optical mapping to correct scaffold arrangement in the tomato genome. G3 4:1395–1405

2. Marx V (2013) The genome jigsaw. Nature 501:263–268

3. The Arabidopsis Genome Initiative (2000) Analysis of the genome sequence of the flowering plant Arabidopsis thaliana. Nature 408:796–815

4. Yu J, Hu S, Wang J, Wong GK, Li S, Liu B, Deng Y, Dai L, Zhou Y, Zhang X et al (2002) A draft sequence of the rice genome (Oryza sativa L. ssp. indica). Science 296:79–92

5. Goff SA, Ricke D, Lan TH, Presting G, Wang RL, Dunn M, Glazebrook J, Sessions A, Oeller P, Varma H et al (2002) A draft sequence of the rice genome (Oryza sativa L. ssp. japonica). Science 296:92–100

6. Choulet F, Alberti A, Theil S, Glover N, Barbe V, Daron J, Pingault L, Sourdille P, Couloux A, Paux E et al (2014) Structural and functional partitioning of bread wheat chromosome 3B. Science 345:1249721

7. Peterson DG, Tomkins JP, Frisch DA, Wing RA, Paterson AH (2000) Construction of plant bacterial artificial chromosome (BAC) libraries in plants: an illustrated guide. J Agric Genom 5, http://www.ncgr.org./research/jag

8. Zhang HB, Woo SS, Wing RA (1996) BAC, YAC and cosmid library construction. In: Foster GD, Twell D (eds) Plant gene isolation: principles and practise. Wiley, New York, pp 75–99

9. Zhang HB, Zhao XP, Ding XL, Paterson AH, Wing RA (1995) Preparation of megabase-size DNA from plant nuclei. Plant J 7:175–184

10. Šimková H, Číhalíková J, Vrána J, Lysák MA, Doležel J (2003) Preparation of HMW DNA from plant nuclei and chromosomes isolated from root tips. Biol Plant 46:369–373

11. Šafář J, Noa-Carrazana JC, Vrána J, Bartoš J, Alkhimova O, Lheureux F, Šimková H, Caruana ML, Doležel J, Piffanelli P (2004) Creation of a BAC resource to study the structure and evolution of the banana (Musa balbisiana) genome. Genome 47:1182–1191

12. Barghini E, Natali L, Giordani T, Cossu RM, Scalabrin S, Cattonaro F, Šimková H, Vrána J, Doležel J, Morgante M, Cavallini A (2014) LTR retrotransposon dynamics in the evolution of the olive (Olea europaea) genome. DNA Res. doi:10.1093/dnares/dsu042

13. Allouis S, Moore G, Bellec A, Sharp R, Faivre-Rampant P, Mortimer K, Pateyron S, Foote TN, Griffiths S, Caboche M, Chalhoub B (2003) Construction and characterization of a hexaploid wheat (Triticum aestivum L.) BAC library from the reference germplasm 'Chinese Spring'. Cereal Res Commun 31:331–338

14. Šafář J, Šimková H, Kubaláková M, Číhalíková J, Suchánková P, Bartoš J, Doležel J (2010) Development of chromosome-specific BAC resources for genomics of bread wheat. Cytogenet Genome Res 129:211–223

15. Doležel J, Šimková H, Šafář J, Kubaláková M, Vrána J, Číhalíková J, Šperková R, Kianian SF, Sourdille P, Lukaszewski AJ, Endo TR, Gill BS (2014) International wheat genome sequencing consortium. A complete set of chromosome BAC libraries for genomics of wheat (Triticum aestivum). In: Abstracts of the International conference "Plant and Animal genome XXII", Sherago International Inc., San Diego, p. 206

16. Mago R, Tabe L, Vautrin S, Šimková H, Kubaláková M, Upadhyaya N, Berges H, Kong X, Breen J, Dolezel J, Appels R, Ellis J, Spielmeyer W (2014) Major haplotype divergence including multiple germin-like protein genes, at the wheat Sr2 adult plant stem rust resistance locus. BMC Plant Biol 14:379

17. Janda J, Šafář J, Kubaláková M, Bartoš J, Kovářová P, Suchánková P, Pateyron S, Číhalíková J, Sourdille P, Šimková H, Fairaivre-Rampant P, Hřibová E, Bernard M, Lukaszewski A, Doležel J, Chalhoub B (2006) Advanced resources for plant genomics: BAC library specific for the short arm of wheat chromosome 1B. Plant J 47:977–986

18. Šimková H, Šafář J, Suchánková P, Kovářová P, Bartoš J, Kubaláková M, Janda J, Číhalíková J, Mago R, Lelley T, Doležel J (2008) A novel resource for genomics of Triticeae: BAC library specific for the short arm of rye (Secale cereale L.) chromosome 1R (1RS). BMC Genomics 9:237

19. Sambrook J, Fritsch E, Maniatis T (1989) Molecular cloning: a laboratory manual, 2nd edn. Cold Spring Harbor Laboratory Press, Cold Spring Habor, NY

# Chapter 12

# The Chromosome Microdissection and Microcloning Technique

## Ying-Xin Zhang, Chuan-Liang Deng, and Zan-Min Hu

## Abstract

Chromosome microdissection followed by microcloning is an efficient tool combining cytogenetics and molecular genetics that can be used for the construction of the high density molecular marker linkage map and fine physical map, the generation of probes for chromosome painting, and the localization and cloning of important genes. Here, we describe a modified technique to microdissect a single chromosome, paint individual chromosomes, and construct single-chromosome DNA libraries.

**Key words** Chromosome microdissection and microcloning, Degenerated oligonucleotide-primed-PCR, Fluorescence in situ hybridization, Single-chromosome library

## 1 Introduction

Since its inception, the chromosome microdissection and microcloning technique [1] has become an efficient and direct approach for isolating DNA from specific chromosomes and/or specific chromosome sections. The isolated DNA is used for genomic research including: (1) genetic linkage map and physical map construction [2, 3], (2) generation of probes for chromosome painting [4–9], and (3) generation of chromosome-specific expressed sequence tags libraries [10–12]. Here, we present a comprehensive protocol to isolate DNA sequences derived from a single chromosome, for which 1R chromosome of rye (L. *Secalecereale*) was used as a model. The protocol includes a technique to prepare metaphase chromosomes from germinating seeds. Such chromosome samples can be used for chromosome microdissection and fluorescence in situ-hybridization mapping. The details on chromosome dissection, DNA amplification, and library construction of microdissected chromosome are provided. Here, we also demonstrate some applications of the dissected chromosome DNA, especially a modified FISH technique to paint individual chromosomes [8, 9, 13].

Shahryar F. Kianian and Penny M.A. Kianian (eds.), *Plant Cytogenetics: Methods and Protocols*,
Methods in Molecular Biology, vol. 1429, DOI 10.1007/978-1-4939-3622-9_12,
© Springer Science+Business Media New York 2016

## 2  Materials

Prepare all solutions using deionized water and analytical grade reagents. Prepare and store all reagents at room temperature (unless indicated otherwise).

### 2.1  Chromosome Spread Preparation

1. 90 % acetic acid: Dilute concentrated acetic acid in water.

2. 70 % ethanol.

3. 5× Citric buffer: 50 mM sodium citrate, 50 mM EDTA, adjust pH to 5.5 by citric acid.

4. Enzyme mixture solution: 1 % pectolyase Y23 and 2 % cellulose "Onozuka" R-10 in 1× citric buffer.

5. Inverted phase-contrast microscope with a micromanipulator.

### 2.2  DOP-PCR (Degenerated Oligonucleotide-Primed-PCR)

1. Proteinase K solution: For 500 μL proteinase solution, dilute 15 μL of 19 mg/mL proteinase K in 485 μL of 1× *Taq* DNA polymerase buffer, split into 10 μL aliquots, and store at –20 °C.

2. DOP-PCR primer:   5′-CCGACTCGAGNNNNNNAT GTGG-3′.

### 2.3  FISH (Fluorescence In Situ Hybridization)

1. 2× DNase buffer: 100 mM Tris–HCl pH 7.5, 10 mM MgCl$_2$, 2 mM 2-mercaptoethanol, 200 μg/mL bovine serum albumin, 0.2 mM phenylmethyl sulfonyl fluoride (optional). Store at –20 °C.

2. DNaseI stock: 1× DNase buffer, 50 % glycerol, DNase I 10 U/μL. Store at –20 °C.

3. DNase I working stock: Dilute the DNase I stock (10 U/μL) to 100 mU/μL concentration with 1× DNase buffer (50 % glycerol).

4. 20× SSC: 3 M NaCl, 0.3 M Sodium Citrate, pH 7.0.

5. Hybridization solution: 2× SSC, 1× TE (pH 7.0).

6. dNTP (-dCTP) mix: 2 mM each of dATP, dGTP, and dTTP.

7. dNTP (-dTTP) mix: 2 mM each of dATP, dGTP, and dCTP.

8. Fluorescence microscope.

### 2.4  Reagents

1. Nitrous oxide (N$_2$O) gas.

2. *Taq* polymerase.

3. 1× *Taq* DNA polymerase buffer.

4. DNA polymerase I.

5. dNTP Solution: 2.5 mM dATP, 2.5 mM dCTP, 2.5 mM dGTP, and 2.5 mM dTTP.

6. Texas Red-5-dCTP.

7. Tide Alexa Fluor 488-5-dUTP.

8. 4′,6-diamidino-2-phenylindole (DAPI).

9. Repetitive sequence specific to chromosome of interest (e.g., pScl19.2 of rye, *see* **Note 1**).

10. Autoclaved salmon sperm DNA (10 mg/mL).

11. Bacteria cloning plasmid (e.g., pMD18-T Vector).

12. DIG High Prime DNA Labeling and Detection Starter Kit.

# 3 Methods

Carry out all procedures on ice unless otherwise specified.

### 3.1 Material Fixation

1. Seeds are germinated on moist filter paper in a petri dish at 25 °C for 2–3 days (*see* **Note 2**).

2. Collect root tips of 1–5 cm in length and place in 1.5 mL centrifuge tube with a hole in the cap for gas treatment (*see* **Note 3**).

3. Treat the collected root tips with nitrous oxide gas at 1.0 MPa for 2 h.

4. Fix the root tips in ice cold 90% acetic acid for 10 min and keep in 70% ethanol at −20 °C for storage (*see* **Note 4**).

### 3.2 Preparation of Chromosome Samples

1. Wash the root tips in ice cold water for 10 min.

2. Pick up the root tips with a forceps and remove the water by rolling them on a piece of dry filter paper, without letting them dry.

3. Cut off the section of the root tips undergoing active cell division on a glass slide and collect them into a 1.5 mL microcentrifuge tube with 20 µL of enzyme mixture (*see* **Note 5**).

4. Incubate the root tips at 37 °C for 30–60 min.

5. After digestion, fill the tubes with 70% ethanol.

6. Carefully remove the 70% ethanol, leaving 20 µL and root tips inside the tube.

7. Break the root tips inside the tube using a dissecting needle until the tissue becomes almost invisible (*see* **Note 6**).

8. Centrifuge briefly in a microcentrifuge, and collect the pellet (*see* **Note 7**).

9. Wash the pellet by resuspending it with 100% ethanol, centrifuge briefly in microcentrifuge, and remove the ethanol (*see* **Note 8**).

10. Add 30 μL of freshly prepared 90% acetic acid-10% methanol mix to resuspend the pellet by vortexing briefly (*see* **Note 9**).

11. Drop the cell suspension onto glass slides kept in a humid chamber (6–10 μL per slides) or keep the cell suspension at –20 °C.

12. Check the slides by a traditional light microscope and select good spreads for chromosome isolation and FISH (*see* **Note 10**).

### 3.3 Microdissection of Target Chromosome

1. Target the chromosome being dissected under an inverted phase-contrast microscope equipped with a micromanipulator.

2. Microdissect the target chromosome using a fine glass needle fixed on the micromanipulator and transfer to a microcentrifuge tube containing 10 μL proteinase K solution (*see* **Note 11**).

3. Centrifuge shortly in microcentrifuge and digest the isolated chromosome with proteinase K at 37 °C for 2 h.

4. Inactive the proteinase K by incubating the tube at 90 °C for 10 min.

### 3.4 DOP-PCR Amplification of the Target Chromosome DNA

1. The first round of PCR amplification is in a 50 μL reaction mixture containing 5 μL of 10× PCR buffer, 1 μL of dNTPs (2.5 mM), 2 μL of DOP-PCR primer (10 μM), 0.5 μL of *Taq* polymerase (10 U/μL), 10 μL of the digestion product, and 31.5 μL of ddH$_2$O. Followed by the PCR program: 94 °C for 5 min, 4 cycles of 1 min at 94 °C, 90 s at 30 °C, 3 min at 72 °C; 24 cycles of 1 min at 94 °C, 1 min at 55 °C, 90 s at 72 °C, and a final extension at 72 °C for 10 min.

2. The second round of PCR reagents are added to a final volume of 50 μL: 5 μL of the first round of PCR products, 5 μL of 10× PCR buffer, 2 μL of DOP-PCR primer (10 μM), 1 μL of dNTPs (2.5 mM), 0.5 μL of *Taq* polymerase (10 U/μL), and 36.5 μL of ddH$_2$O. The PCR program is as follows: 5 min at 94 °C, 34 cycles of 1 min at 94 °C, 90 s at 50 °C, 3 min at 72 °C, and a final extension at 72 °C for 5 min (*see* **Note 12**).

3. Check the second round of PCR products by running a 1% agarose gel (*see* **Note 13**).

### 3.5 Probe Labeling of the Target Chromosomes DNA

1. After the second round of PCR, the product is directly used for probe labeling via nick translation reaction: 10 μL of the second round PCR product, 2 μL of nick translation buffer, 2 μL of dNTP (-dCTP) mix, 0.5 μL of Texas Red-5-dCTP (1 mM), 0.5 μL of DNase I (100 mU/μL), 5 μL of DNA polymerase I (10 U/μL). 2 μg of the repetitive sequence pSc119.2

was labeled with Tide Alexa Fluor 488-5-dUTP (1 mM), using 2 µL of dNTP (-dTTP) mix instead of dNTP (-dCTP) mix. The labeling of both probes is taking place in two different tubes.

2. Incubate the tube at 15 °C for 2 h.

3. Add 30 µg autoclaved salmon sperm DNA (10 mg/mL), vortex.

4. Add 2.5 volumes of 90% ethanol-10% sodium acetate mix (3 M, pH 5.2), vortex, store the tube at –20 °C for 2 h or overnight.

5. Centrifuge at $16,200 \times g$ in a microcentrifuge at 4 °C for 30 min, and decant the supernatant.

6. Wash the pellet with 70% ethanol, remove ethanol, and dry.

7. Dissolve the pellet with 20 µL of hybridization solution and keep in the dark at –20 °C.

**3.6 Characterization of DNA from Microdissected Chromosome by FISH**

1. Cross-link root tip chromatin to slides by exposure to UV light, 120–125 mJ per square (*see* **Note 14**).

2. Dilute 0.5 µL of probe with 5 µL of hybridization buffer.

3. Apply an aliquot of 5.5 µL of the diluted probe to the prepared metaphase cells and place a $22 \times 22$ mm plastic coverslip over the area of cells.

4. Place the slides on wet tissues in the bottom of a water tight box and then place the box into boiling water for 5 min to denature root tip DNA and the probe DNA (on slides) simultaneously.

5. Take out the slides and transfer them into a preheated humid storage container lined with Kimwipes moistened by 2× SSC, hybridization is performed at 55 °C overnight.

6. Wash the slides in 2× SSC to remove coverslips and excess probe.

7. Apply one drop of DAPI (1.5 µg/mL) to the metaphase cells and capture the images with a camera under a fluorescence microscope (*see* **Note 15**).

**3.7 Construction of Single-Chromosome Library**

1. After DOP-PCR, 4 µL of the second round of PCR products were ligated to pMD18-T Vector for transformation to generate a library of the microdissected chromosome.

2. Transformed bacterial colonies can be checked for microdissected chromosome segment and copy number using PCR and dot blot hybridization before sequencing (*see* **Notes 16 and 17**).

## 4  Notes

1. Repetitive sequence pSc119.2: kindly provided by Professor Fang-Pu Han (Institute of Genetics and Developmental Biology, Chinese Academy of Sciences).

2. Different plant materials have unique optimum germination temperatures; for wheat, 25 °C is suitable.

3. Place no more than 11 root tips in one tube.

4. Fix root tips in ice cold 90% acetic acid (should be no longer than 1 h) and the fixed root tip can be kept for years in 70% ethanol at –20 °C.

5. This step can be performed on a dark surface to facilitate the identification of the actively dividing region of root tips.

6. If the root tips are hard to break, the digestion time in **step 4** should be longer, if the root tip is almost invisible before break, the digestion time should be shortened.

7. Centrifuge 10–20 s in a microcentrifuge, do not exceed $845 \times g$.

8. Tap the tube with fingers or vortex shortly to suspend the cells.

9. Before dropping the cell suspension onto slides, keep tubes containing cells on ice. 100% acetic acid can also be used, but cells must be dropped immediately after leaving the ice or acetic acid.

10. Slide for chromosome microdissection should be used immediately. It is not desirable to store slides. For FISH, the slide can be stored for years at –20 °C.

11. Before dissecting the chromosome, 0.5 μL of 50% ethanol should be dropped onto the target chromosome to make it easier to isolate the chromosome. If the chromosome was successfully dissected, it should adhere to the tip of the glass needle, which would be observed under the microscope (Fig. 1).

12. A negative control (no template DNA) and a positive control (genomic DNA as template) should be conducted throughout the whole DOP-PCR process.

13. After two successive rounds of amplification, DNA products obtained from the dissected chromosome can be observed as bright smears ranging in size from 250 to 2000 bp. The positive control of genomic DNA should also show bright smears with a wider range of fragment sizes, and no products should be observed from the negative control (Fig. 2).

14. Cross-linked slides should be used immediately for hybridization or be stored at –20 °C.

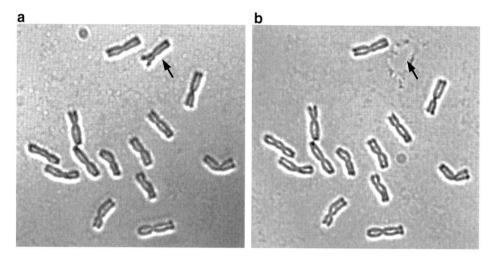

**Fig. 1** Microdissection of 1R chromosome of rye using a micromanipulator. (**a**) Mitotic metaphase chromosome of root tip cells before chromosome microdissection. (**b**) Metaphase chromosome after chromosome microdissection. The *arrow* indicates 1R chromosome. Bar = 10 μm

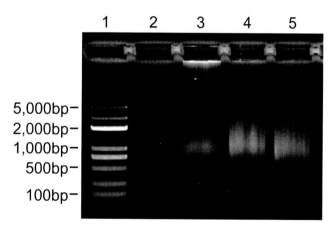

**Fig. 2** DOP-PCR of microdissected chromosome. (*Lane 1*) DNA molecular weight marker, (*lane 2*) negative control (PCR product without DNA template), (*lane 3*) positive control (PCR product with rye genomic DNA as the template), and (*lanes 4* and *5*) DOP-PCR products from microdissected chromosomes

15. The characterization of amplified DNA from microdissected chromosome is important for single-chromosome microdissection and microcloning. Here, we use fluorescence in situ hybridization to ascertain the origin of amplified DNA (Fig. 3). Any DNA contamination from other chromosome should be avoided, however, because repetitive sequences account for a large portion of the plant genomes (e.g., up to 92% of rye (*Secalecereale* L.) genome), the amplified DNAs

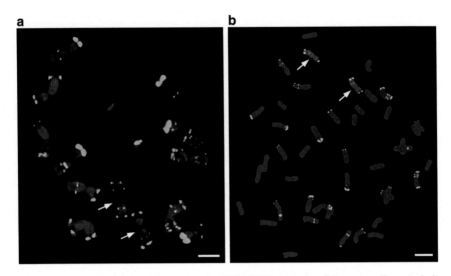

**Fig. 3** FISH patterns on mitotic metaphase using the DOP-PCR products of the microdissected chromosome (*red*) and pSc119.2 (*green*) as probes (this probe can be used for the identification of individual rye chromosome). (**a**) FISH patterns on *Secale cereale* L. var. KingII and (**b**) FISH patterns on wheat-rye 1R chromosome addition line. The *arrows* indicate 1R chromosomes. Bar = 10 μm

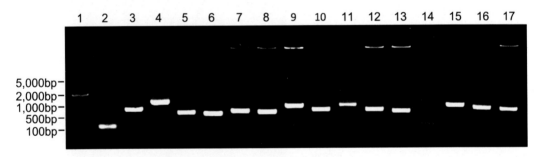

**Fig. 4** A 1 % agarose gel showing amplified insert sequences from the library of microdissected chromosome. (*Lane 1*) 5000 bp DNA ladder marker and (*lanes 2–17*) the inserts from the selected clones

can only be traced to the genome of that particular species (e.g., rye in this experiment), not the dissected 1R chromosome.

16. About $3 \times 10^6$ recombinant microclones were obtained. Positive recombinant clones containing inserts of the lengths 500–2000 bp were preselected on the basis of 1 % agarose gel electrophoresis (Fig. 4). Subsequent selection was performed by dot blot using DIG-labeled rye genomic DNA.

17. According to the intensity of hybridization signals, the low copy or unique copy DNA sequences in genomes should show weak or no signals, while the medium or high copy DNA sequences show strong signals (Fig. 5).

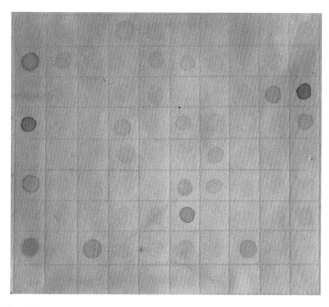

**Fig. 5** Dot blot screening of single-chromosome DNA library using DIG-labeled *Secale cereale* L. var. KingII genomic DNA as the probe. Strong signals represent the medium or high copy DNA sequences in the rye genome, while weak signals represent the low copy DNA sequences

## Acknowledgments

This work was supported by the National Natural Science Foundation of China (31170209) and the Ministry of Science and Technology of China (2011AA100101). We thank Dr. Fangpu Han for kindly providing repetitive sequence pSc119.2 and the technique advice on FISH.

## References

1. Scalenghe F, Turco E, Edström JE, Pirrotta V, Melli M (1981) Microdissection and cloning of DNA from a specific region of *Drosophila melanogaster* polytene chromosomes. Chromosoma 82:205–216

2. Jiang CZ, Song WQ, Li XL, Chen RY (1998) Studies on microdissection and microcloning of the rye chromosome 1R. Acta Bot Sin 40:988–993

3. Seifertova E, Zimmerman LB, Gilchrist MJ, Macha J, Kubickova S, Cernohorska H, Zarsky V, Owens NDL, Sesay AK, Tlapakova T, Krylov V (2013) Efficient high-throughput sequencing of a laser microdissected chromosome arm. BMC Genomics 14:357

4. Thalhammer S, Langer S, Speicher MR, Heckl WM, Geigl JB (2004) Generation of chromosome painting probes from single chromosomes by laser microdissection and linker-adaptor PCR. Chromosome Res 12:337–343

5. Nie WH, Wang JH, Su WT, Wang YX, Yang FT (2009) Chromosomal rearrangements underlying karyotype differences between Chinese pangolin (*Manis pentadactyla*) and Malayan pangolin (*Manis javanica*) revealed by chromosome painting. Chromosome Res 17:321–329

6. Nie W, Wang J, Su W, Wang D, Tanomtong A, Perelman PL, Graphodatsky AS, Yang F (2012)

Chromosomal rearrangements and karyotype evolution in carnivores revealed by chromosome painting. Heredity 108:17–27

7. Pauciullo A, Nicodemo D, Cosenza G, Peretti V, Iannuzzi A, Di Meo GP, Ramunno L, Iannuzzi L, Rubes J, Di Berardino D (2012) Similar rates of chromosomal aberrant secondary oocytes in two indigenous cattle (*Bostaurus*) breeds as determined by dual-color FISH. Theriogenology 77:675–683

8. Deng CL, Bai LL, Fu SL, Yin WB, Zhang YX, Chen YH, Wang RRC, Zhang XQ, Han FP, Hu ZM (2013) Microdissection and chromosome painting of the alien chromosome in an addition line of wheat - *Thinopyrum intermedium*. PLoS One 8:e72564

9. Deng CL, Bai LL, Li SF, Zhang YX, Li X, Chen YH, Wang RRC, Han FP, Hu ZM (2014) DOP-PCR based painting of rye chromosomes in a wheat background. Genome 57:473–479

10. Zhou RN, Hu ZM (2007) The development of chromosome microdissection and microcloning technique and its applications in genomic research. Curr Genomics 8:67–72

11. Zhou RN, Shi R, Jiang SM, Yin WB, Wang HH, Chen YH, Hu J, Wang RRC, Zhang XQ, Hu ZM (2008) Rapid EST isolation from chromosome 1R of rye. BMC Plant Biol 8:28

12. Jiang SM, Yin WB, Hu J, Shi R, Zhou RN, Chen YH, Zhou GH, Wang RRC, Song LY, Hu ZM (2009) Isolation of expressed sequences from a specific chromosome of *Thinopyrum intermedium* infected by BYDV. Genome 52:68–76

13. Han FP, Gao Z, Birchler JA (2009) Reactivation of an inactive centromere reveals epigenetic and structural components for centromere specification in maize. Plant Cell 21:1929–1939

# Chapter 13

## Immunolocalization on Whole Anther Chromosome Spreads for Male Meiosis

### Stefanie Dukowic-Schulze, Anthony Harris, and Changbin Chen

### Abstract

Immunolocalization of cells undergoing meiosis has proven to be one of the most important tools to decipher chromatin-associated protein dynamics and causal relationships. Here, we describe a protocol established for maize which is easily adaptable to other plants, for example, with minor modifications to *Arabidopsis* as stated here. In contrast to many other protocols, the following protocol is based on fixation by a 3:1 mixture of ethanol and acetic acid. Spreading of cells is followed by freeze-shattering, protein antigenicity retrieval by a hot citrate buffer bath, antibody incubations and washes, and DNA staining.

**Key words** Immunolocalization, Meiosis, Maize, Arabidopsis, Plants, Anthers, Chromosome spread

## 1 Introduction

Immunodetection of proteins is a widely used method to gain knowledge about spatial and temporal proceedings in the cell and can be performed on whole tissue sections or on squashed cells [1]. Immunolocalization techniques are especially valuable for examining chromosome-associated proteins during meiosis, signified by ongoing findings on histone and nonhistone proteins (*reviewed in* [2]). Most importantly, insight into the distribution of recombination-associated proteins during prophase I of meiosis has supported the building of present models regarding initiation and progression of pairing, synapsis, and recombination [3–7].

Recently, many protocols on plant meiosis immunolocalization have been published, apparently yielding excellent results (e.g. [8–12]). Most of them are dedicated to the dicot model plant *Arabidopsis* and are based on formaldehyde fixation. Formaldehyde fixation strengthens connections between proteins and nucleic acids, but can occasionally have a disadvantage by masking antigenicity. Preparations on plant material initially fixed in mixtures of ethanol and acetic acid are also common for nonmembrane-associated

Shahryar F. Kianian and Penny M.A. Kianian (eds.), *Plant Cytogenetics: Methods and Protocols*,
Methods in Molecular Biology, vol. 1429, DOI 10.1007/978-1-4939-3622-9_13,
© Springer Science+Business Media New York 2016

proteins, and frequently followed up with a short formaldehyde postfixation step before antibody is applied. The spreading procedure is the most crucial but also the most challenging part for obtaining high-quality results. In our hands, spreading and freeze-shattering of maize or *Arabidopsis* cells fixed in ethanol and acetic acid are the superior method when compared with other approaches.

Though most protocols end with microscopy, this protocol provides support for the nontrivial downstream processing of images and how to achieve publication-worthy quality. Described are some helpful step-by-step procedures using ImageJ to edit and polish images and offer general advice on the best practices and ethical principles to achieve responsible conduct of research when processing images [13]. The following protocol describes step-by-step the procedure of sampling, fixation, chromosome spreading, antigen retrieval, binding of antibodies, and subsequent detection by conjugated fluorochromes for meiosis samples, concluding with image processing (*see* **Note 1**).

## 2   Materials

All solutions are prepared using double deionized water. If is not otherwise indicated, solutions, reagents, and buffers used in this protocol are prepared and stored at room temperature.

### 2.1   Plant Material

Plants can be grown in the field, well-controlled greenhouse, or growth chamber.

### 2.2   Equipment

1. Fluorescence microscope with adequate filters.

2. Slide warmer at 42 °C.

3. Microscope slides with improved adherence (can be purchased or self-coated with polylysine: incubate slides in 0.01 % polylysine for 15 min on each side, let air-dry afterwards).

4. Coplin jars, staining dishes.

5. Magnetic stirrer with heat element.

6. Dissecting instruments: Forceps, razor blade or scalpel, 1 ml disposable syringes with 27 guage sized needle.

7. Nail polish or professional slide sealer.

8. Diamond pen.

### 2.3   Self-Made Utensils

1. Disposable plastic pestles: Heat the thin end of a 200 μl pipette tip by holding it just above a flame. When the plastic starts melting, quickly press it down on a glass microscope slide so that it forms a round platform of up to 5 mm diameter. While cooling, it changes the color from transparent to whitish (Fig. 1a).

2. Oxidized iron rods (i.e. rusty nails): Roughen the surface of a couple nails with sandpaper (especially at the bottom part), dip

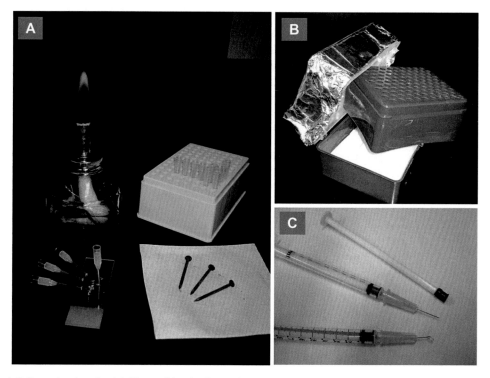

**Fig. 1** Self-made utensils. (**a**) Disposable plastic pestles, made out of pipette tips, using a flame and a glass slide; oxidized iron rods (rusty nails). (**b**) Humidity chamber/box for slide incubation at higher temperature or for longer duration. (**c**) Syringe needles for dissection, syringe plunger for squashing, and needle bent into hook for gentle poking

      them into water, and let them sit outside (e.g. in the lid of a glass jar) until they develop rust (Fig. 1a).

3. Humidity chamber: Cover any transparent parts of an unused 1000 ml pipette tip box with tin foil to protect light-sensitive samples. Temporarily remove the tip box insert with holes to add some layers of tissue wipes, moisten with dH$_2$O, put the insert back in (Fig. 1b).

4. Syringe needle with hook: Bend the thin needle to form a hook by pressing the needle on a hard surface or using forceps (Fig. 1c).

5. Parafilm coverslips: Cut to pieces of ~24 × 30 mm.

*2.4  Buffers, Solutions, and Chemicals*

1. Farmer's fixative (three parts ethanol, one part acetic acid): Prepare 500 ml consisting of 375 ml 100 % ethanol and 125 ml glacial acetic acid.

2. Acetocarmine staining solution: Use a magnetic stirrer with an integrated heat element, put 0.5 g carmine in 55 ml water, while stirring add 45 ml glacial acetic acid, then boil for around 30 min. Let the solution cool down afterwards and filter it through a Whatman paper into a bottle with tinted glass or tin foil wrapped around. Store in the dark.

3. 10 mM citrate buffers (pH 4.5 for initial washes, pH 6.0 for antigen retrieval): Prepare 0.1 M stock solutions of citric acid (21.014 g citric acid monohydrate in 1 l dH$_2$O), and of sodium citrate (29.41 g (tri-)sodium citrate dihydrate in 1 l dH$_2$O). For 50 ml 10 mM citrate buffer pH 4.5, mix 2.6 ml 0.1 M citric acid, 2.4 ml 0.1 M sodium citrate, and 45 ml dH$_2$O. For 500 ml 10 mM citrate buffer pH 6.0, mix 5.75 ml 0.1 M citric acid with 44.25 ml 0.1 M sodium citrate and 450 ml water.

4. Enzyme stock solutions: Prepare 3% enzyme stock solutions each for cellulose, pectolyase, and cytohelicase, e.g. 0.3 g in 10 ml 10 mM citrate buffer pH 4.5. Freeze 1 ml aliquots at −20 °C.

5. Digest mixture A: To make the working solution of digest mixture A with 0.3% of each enzyme, use 1 ml of each cellulose, pectolyase and cytohelicase stock, and add 7 ml 10 mM citrate buffer pH 4.5. Store aliquots at −20 °C or leave at 4 °C (*see* **Note 2**).

6. Digest mixture B: A mixture of 0.4% cytohelicase, 1% polyvinylpyrrolidone (PVP), 1.5% sucrose in water. For 5 ml digest mixture B, we use 666 µl aliquot 3% cytohelicase, 0.5 ml PVP and 0.75 g sucrose, add dH$_2$O to 5 ml. Store aliquots at −20 °C or leave at 4 °C (*see* **Note 2**).

7. 60% acetic acid: Mix 50 ml by slowly adding 30 ml glacial acetic acid to 20 ml dH$_2$O.

8. Dry ice pellets.

9. 1× PBS: commercial or self-made, best from 10× PBS stock (1 l with 80 g NaCl, 2 g KCl, 14.4 g Na$_2$HPO$_4$, 2.4 g KH$_2$PO$_4$, adjusted to pH 7.4 with HCl).

10. PBST: 1× PBS with 0.1% Triton-X 100.

11. DAPI solution: 10 µg/ml DAPI (4′,6-diamidino-2-phenylindole) in 1× PBS.

12. VectaShield or other fluorescence-preserving mounting medium.

**2.5 Antibody Solutions**

1. Antibody dilution buffer: PBST (1× PBS with 0.1% Triton-X) with 1% BSA.

2. Primary antibody.

3. Secondary fluorophore-conjugated antibody.

# 3   Methods

The main method described here is for a quick protocol using whole anthers from maize (*Zea mays*). In addition, we provide information on how to easily adapt the protocol for *Arabidopsis* anthers in Table 1.

**Table 1**

**Adjustments for Arabidopsis**

| Section | Protocol modifications |
| --- | --- |
| 3.1 | Instead of spikelets, use whole perfect inflorescences. Fix inflorescences as in **steps 3** till **5** of the section. Small microtubes can be used. |
| 3.2 | Remove all open flowers and most buds longer than 0.5 mm. Omit **step 1**, then treat the remaining inflorescences as in **step 2–7** but preferably use digest mixture A and extend the digest step at **step 5** to 30 min as stated in other works for chromosome spreads for *Arabidopsis*. |
| 3.3 | Put an inflorescence on a glass slide or a petri dish with wet filter paper. Using a binocular, dissect the anthers out of buds smaller than 0.5 mm (*see* **Note 3**). Place the anthers in 10 µl of 60 % acetic acid sitting on a microscope slide. Then continue as described in **steps 2** till **5**. |
| 3.4–3.6 | *As described* |

**3.1 Plant Material Preparation for Maize**

1. Collect florets of interest undergoing meiosis, place them in a container with Farmer's fixative (*see* **Note 4**).

2. Determine the meiotic stage by acetocarmine staining (*see* **Note 5**). Dissect the anthers from a spikelet and transfer them onto a glass slide with few drops of acetocarmine solution (~50 µl). Heat the sample for a few seconds over a flame, such as from an ethanol burner, then use an oxidized iron rod (i.e. a rusty nail) to stir, surrounding the sample with circular movements, mixing the acetocarmine until it turns into a brownish or purplish color. Heat the slide again for a few seconds, then put it down and smash the anthers with a disposable pipette tip pestle. Add a coverslip; press it down with a tissue wipe underneath and on top to drain excessive stain away. Examine with a bright field microscope to determine the meiotic stage (*see* **Note 6**).

3. Once the proper meiotic stage is identified from dissected anthers, fix all corresponding male reproductive tissue in Farmer's fixative. Large samples, such as spikelets from a whole maize tassel can be put in a 50 ml tube with at least 20× volume (when compared to the volume occupied by the sample) of Farmer's fixative (*see* **Note 7**).

4. Replace the fixative once after a couple hours or overnight incubation with Farmer's fixative.

5. Fixed spikelets can be stored in the fixative for several months at −20 °C.

**3.2  Initial Sample Treatment**

1. On a clean microscope slide, dissect anthers using disposable syringe needles and a dissecting microscope with ~10× magnification (*see* **Note 8**). Quickly transfer the anthers into sufficient citrate buffer pH 4.5, e.g. 200 μl in a well of a 96-well plate for 3–9 anthers.

2. Wash fixed anthers by incubating in citrate buffer for 5–10 min (*see* **Note 9**).

3. Replace the citrate buffer and incubate for another 5–10 min.

4. Remove the citrate buffer completely with a 200 μl pipette tip, tilting the 96-well plate to gather excessive buffer at one side of the well.

5. Add 20–50 μl of digest mixture A or B, making sure to cover all anthers completely.

6. Incubate for 10 min at 37 °C (*see* **Note 10**).

7. Stop the digestion by putting the 96-well plate on ice and add 200 μl of citrate buffer. Promptly remove all liquid and replace with 200 μl of citrate buffer to dilute any residual enzyme.

**3.3  Chromosome Spread**

1. Transfer 2–5 fixed anthers onto a microscope slide with 10 μl of 60% acetic acid. Squash them using a self-made pipette tip pestle (Fig. 2a, b).

2. Add a second slide perpendicular on top (adhesive side down), then fix the position of the slides by firmly holding them down (Fig. 2c) and thoroughly tapping onto the slide layers with e.g. a plunger (the inlet of a syringe with a rubber end) (*see* **Note 11**). Then apply additional even pressure on the slide layers, using a thumb to press down till your nail area becomes whitish (*see* **Note 12**).

3. Put the perpendicular crossed slides onto aligned uniform dry ice pellets and press them gently down, wearing (cloth) gloves or using a crumpled-up tissue wipe to protect your fingertips from the cold. Press down for approximately 20 s, then slowly let go, and put two dry ice pellets on top of the slide-cross (Fig. 2d, e).

4. Wait for 5 min before breaking the frozen slide-cross apart, holding the slides close to the joint area while doing so (Fig. 2f). Put both slides face-up onto a slide warmer at 42 °C for at least 30 min until all liquid is gone.

5. After drying, use a diamond pen to draw a circle around the sample area (*see* **Note 13**).

**3.4  Antigen Reactivation/Retrieval**

1. Heat ~250 ml of citrate buffer in an unused 1000 ml pipette tip box in the microwave until it boils. Leaving the box in the microwave, submerge a slide rack with the slides to incubate for 5 min, heating it to boiling once more (*see* **Note 14**).

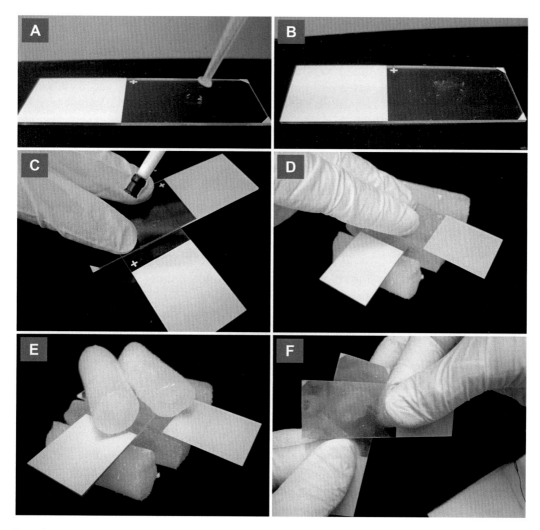

**Fig. 2** Spreading and freeze-fracturing procedure. (**a**) Dissected anthers in a drop of 60 % acetic acid, ready to be smashed with a plastic pestle. (**b**) Smashed anthers, stirred and evenly distributed in an area of approximately 1–1.5 cm in diameter. (**c**) Slides perpendicular as a cross, generated by pressing a second slide face-down onto the sample area, ready to be repeatedly tapped with a syringe plunger, and then a final strong push down with a thumb. (**d**) Slides perpendicular as a cross, lightly pressed down on top of dry-ice pellets. (**e**) Dry-ice stack for completely freezing the sample. (**f**) Gentle breaking-apart of the slide cross by tilting one slide against the other, gripping close at the joined area

2. Transfer slides into a Coplin jar or a staining dish with 1× PBS, using a forceps. Incubate for 5–10 min and then transfer slides to a new jar with 1× PBS, incubating again for 5–10 min.

3. Perform the third wash with PBST, pipetting 60 μl onto the sample area indicated by diamond pen marking. Then cover the area with a piece of parafilm (sized ~24 × 30 mm) (*see* **Note 15**). Incubate for at least 10 min, placing the slides into a humidity box in case of prolonged incubation (*see* **Note 16**).

4. Remove the piece of parafilm with a forceps, picking it straight up rather than sliding it off sideways. On lint free tissue wipes (such as Kimwipes), raise one edge of the slides, allowing the liquid to pour down and absorbed. Use a tissue wipe to wipe any liquid away from underneath the slides and to blot away any accumulated liquid still visible around the sample area.

*3.5 Immunolocalization*

1. Add 50 μl of the first antibody solution in the middle of each sample, then use a piece of parafilm to spread the solution and cover the whole sample area, eliminating any air bubbles. Incubate in a humidity chamber at 4 °C overnight.

2. Remove the antibody solution as described in **step 4** of Section 3.4.

3. Wash slide twice for 5–10 min in 1× PBS in a Coplin jar, then another time with 60 μl of PBST directly on the slide. Incubate for 10 min or longer, placing the slides into a humidity box in case of prolonged incubation.

4. Apply the secondary antibody as in **step 1** of Section 3.5, from now on protecting the samples/slides from light (*see* **Note 17**). Start with a dilution of 1:100 or any suggestion by the manufacturer (*see* **Note 18**). Incubate at 37 °C for 1 h in a humidity chamber.

5. Wash three times, as above in **step 2** of Section 3.5.

*3.6 DAPI Counterstaining (See Note 19)*

1. Apply 50 μl DAPI staining solution, cover with a piece of parafilm, and incubate at 37 °C in a humidity box for 15 min or longer.

2. Wash the slides in Coplin jars three times for 5 min each: First in 1× PBS, then twice in distilled water, all in the dark (*see* **Note 20**).

3. Let the slide dry slightly and then add 15 μl of VectaShield or another fluorescence-preserving mounting medium. Apply a coverslip and a thin tissue wipe on top, and then apply gentle pressure to flatten the sample while absorbing excessive solution at the coverslip rims. Ensure the coverslip does not slide when performing this step.

4. To permanently seal the slides, use nail polish, applying it as a thin line covering the transition from coverslip to glass slide all around. Let dry in the dark for at least 15 min. Slides can then be stored at –20 °C.

5. Examine your samples with a fluorescence microscope with the appropriate filters (Table 2). Species with large chromosomes are seen easily at a total magnification of 400×, but image quality is enhanced if an oil-immersed 100× objective is used (*see* **Note 21**).

**Table 2**

**Commonly used fluorescence filters and dyes[a]**

| Fluorophore | Excitation (nm) | Emission (nm) | Example filters |
|---|---|---|---|
| GFP (green fluorescent protein) | ~450–500 | ~500–520 | Green filter (~460–500 nm excitation, ~520–560 nm emission) |
| CFP[b] (cyan fluorescent protein) | ~420–470 | ~470–520 | |
| FITC (fluorescein iso-thiocyanate) | ~480–510 | ~510–540 | |
| Sytox Green[c] | ~480–520 | ~520–550 | |
| DAPI[c] (4′,6-diamidino-2-phenylindole) | ~330–380 | ~430–500 | Blue filter (~325–375 nm excitation, ~435–585 nm emission) |
| Hoechst[c] (33342 or 33258) | ~330–380 | ~420–500 | |
| PI[c] (propidium iodide) | ~490–570 | ~590–660 | Red filter (~515–560 nm excitation, ≥590 nm emission) |
| TRITC (tetramethylrhodamine isothiocyanate) | ~540–560 | ~580–620 | |
| Texas Red[b] | ~570–620 | ~600–650 | |

[a]Exact excitation and emission peaks vary in different solutions; listed are the wavelengths ranges with more than ~60% maximal emission.
[b]Suboptimal fluorophore properties for the example filters listed here.
[c]Fluorophores for staining nucleic acids.

*3.7 Image Processing with ImageJ*

ImageJ [14], also known as Fiji, is an excellent open-source software to edit fluorescent images and create publication-quality images (*see* **Note 22**). Originally developed by the NIH (U.S. National Institutes of Health), it now can be found as Fiji (Fiji is just ImageJ) which contains many useful plugins and has a focus on microscopy image processing.

1. **Basics on using ImageJ**. To open images, simply "drag" and "drop" them in the ImageJ software bar or click "File > Open....". Work only on copies of the original images — some modifications, including adding a scale bar, cannot be undone! Useful tools for initial adjustment can be found under "Image > Adjust > e.g. Brightness/Contrast" (*see* **Note 23**). Special filters are available to remove noise (e.g. "Process > Noise > Despeckle").

2. **Calibration and Scale Bar**. If your imaging software used to capture microscope images not provide information on the scale, first take pictures of a micrometer slide (containing tick marks at specific distances, e.g. 20 μm) at different magnifications, note the pixel size of your images (e.g. 2048 × 2048) (*see* **Note 24**), and measure the known μm distances in pixels. For this, open the image in ImageJ, select the line tool, and

draw a line between two borders, noting the line length shown in the ImageJ bar. Set the scale for one or all open images by "Analyze > Set Scale > …", entering e.g. distance in pixels = 200, known distance = 10, unit of length = μm, check "Global" to apply to all open images. Then add a scale bar by "Analyze > Tools > Scale Bar" and adjust it to your liking (*see* **Note 25**).

3. **Overlaying different focus planes and fluorescence channels**. When working at high magnification, multiple images may be required at slightly different focus planes. To combine them afterwards, generate a stack ("Image > Stacks > Images to Stack") or simply have them in a folder which you drag onto the ImageJ bar. Either have a stack per color to start with or make substacks by "Image > Duplicate >", then entering the slice numbers and a name and color. To improve contrast, first convert to 8-bit image ("Image > Type > 8-bit"), then adjust Brightness/Color and Noise as described before. Then try the different algorithms under "Image > Stacks > Z Project" to obtain the optimal combination (*see* **Note 26**). To merge different fluorescent colors, use "Image > Color > Merge Channels" (*see* **Note 27**).

4. **Preparing Montages**. Multipanel figures of same-sized images for presentations and publications are easily generated by using the crop tool in conjunction with cloned selection shapes. First, open all images you want to crop to the same size. Choose a selection tool (most commonly the rectangle) and draw a boundary around your region of interest in one image. The size of the shape (e.g. $250 \times 250$ pixels) shows up in the ImageJ bar. To add the same-sized selection rectangle in other images: "Edit > Selection > Restore Selection". To crop, click "Image > Crop". The resulting images must be combined in a stack by "Image > Stacks > Images to Stack". Then use "Image > Stacks > Make Montage" where you define the row and column number and add a border line between images. The resulting montage cannot be modified, which is possible by using the Magic Montage Plugin instead (*see* **Note 27**). With this tool, new buttons will appear in the ImageJ bar for montage creation and later rearrangement of the tiles, easy overlay by drag and drop, adding more images (also of different original size), and subsetting is possible (right mouse click on montage). Figures 3 and 4 have been created using the Magic Montage Plugin, showing co-immunolocalizations in maize (Fig. 3) and BrdU immunolocalization used for meiotic time-course determination [15] in Arabidopsis (Fig. 4).

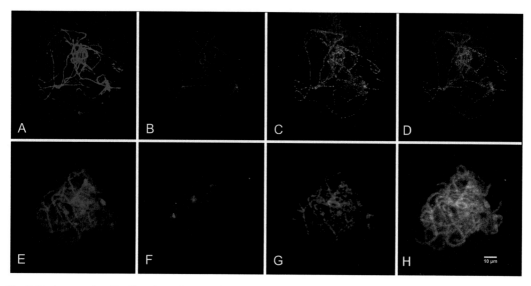

**Fig. 3** Co-immunolocalizations in maize. (**a–d**) Late pachytene. (**a**) DNA (*blue*). (**b**) ZYP1 element of the synaptonemal complex (*red*). (**c**) H3K4me3 (*green*). (**d**) Composite image. (**e–h**) Zygotene-pachytene. (**e**) DNA (*blue*). (**f**) ZYP1 (*red*). (**g**) ASY1 element of the synaptonemal complex (*green*). (**h**) Composite image

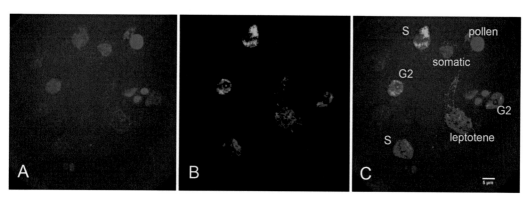

**Fig. 4** BrdU immunolocalization in Arabidopsis. (**a**) DNA stained *blue* by DAPI. (**b**) BrdU in *green*, detected by secondary antibody against primary anti-BrdU antibody. (**c**) Composite image, identity of BrdU-labeled meiotic or premeiotic cells and other nonlabeled cells displayed

## 4  Notes

1. For researchers who are not familiar with chromosome spread and immunolocalization, we recommend to practice chromosome spreads first, omitting Sections 3.4 and 3.5 of the protocol, and checking the spreading results by DAPI staining (Section 3.6).

2. To prevent commonly occurring enzyme mix contamination, filter-sterilize the citrate buffer first and/or keep aliquots at −20 °C. Also, we recommend having separate stock solutions

of the enzymes because their activity can vary and adaptation with different percentages can thus be easily achieved and tried. The enzyme mixes should be stored in the dark.

3. The shape of Arabidopsis buds undergoing male meiosis are rather round than long. Typical syringe needles are ~0.5 mm wide (narrower at the pointed tip), which is a good approximate measure for the desired bud length.

4. Maize plants of most inbred lines we worked with (i.e., B73, Mo17, CML228) are 6–8 weeks old and still contain their tassel (the male inflorescence) inside the stalk when appropriate meiotic stages can be found. To check whether desired meiocytes are present, use a razor blade to make a longitudinal incision, starting a few centimeters above the last node and slicing downwards for around 10 cm, thus slicing through the node. The incision should be aimed to reach the center of the stalk, at first cutting only as deep as one third of the stalk to keep the tassel intact. Gently use your thumbs to spread the layers apart, searching for the tassel inside. The length of the main tassel should be approximately as long as a forefinger (shorter for early meiosis stages), but sometimes only part of the tassel is visible. After removal of sample spikelets, press the stalk material around the incision back together and bandage it tightly with adhesive tape. Since the mid portion of the main tassel has the best synchronized meiocytes which are commonly also the most progressed, taking samples from there is preferred. Using paper-based tape is convenient because it is easy to label the plant on it with the date and a number corresponding to the microtube containing the sample. We often have multiple plants growing and being tested at the same time, and second testing or sampling is thus greatly facilitated.

5. Anthers from the upper floret of a spikelet of maize are bigger and usually farther progressed in meiosis than anthers from the lower floret. For test staining, we recommend staining them separately.

6. For examples of stages of meiosis (*see* ref. 16).

7. Penetration of the fixative into the plant material can be improved by placing the open tubes under vacuum. Complete fixation is achieved when samples sink down to the bottom.

8. We recommend using only the upper (bigger) floret anthers of maize spikelets. Their meiocytes are bigger and more consistency is achieved in this manner. Using both florets gives more different meiotic stages, but we caution about it since the size of anthers and meiotic cells can differ.

9. We find it beneficial to submerge the anthers by dipping them with a needle (formed into a hook to avoid poking holes in the anthers), although this step is optional.

10. Anthers can also be transferred into PCR tubes and incubated in a thermal cycler.

11. It is imperative to avoid any sliding motion of the slides from this point on because this can create artifacts on the chromosome spreads by rupturing and dislocating them.

12. Chromosomes of different species may require different amount of pressure to create best spreads. Grasses, such as maize, generally require more pressure than *Arabidopsis*. Experimentation may be necessary (*see* **Note 1**) to determine the correct pressure before completing the protocol.

13. Though the samples can be stored at this point (best at −20 °C or 4 °C), we get the best results if proceeding right away and incubating with the first antibody overnight. It is an option to first check one of the slides with DAPI staining and then proceed with the corresponding perpendicular slide of confirmed quality i.e. with many and well spread meiocytes. This is especially recommended for beginners.

14. Glass staining dishes are not recommended for heating above 80 °C. A safe alternative is to use commercial microwaveable glass ware or a heat tolerant plastic ware. We did not see a difference regarding the quantity or quality of the following immunostaining between 2, 5, or 10 min.

15. Since Triton-X is classified as dangerous to the environment, we prefer to use a small amount of PBST pipetted onto the slide instead of needing a whole Coplin jar with the solution.

16. Alternatively, all three washes (here and later in the protocol) can be done either in 1× PBS or with detergent. In optimal spreads, no cellular membranes should have to be crossed by the antibodies, thus having no need of detergent though we include it in the antibody solution. Adding detergent can even result in the loss of proteins that are not bound tightly (*see* ref. 17).

17. From this step on, the sample is light-sensitive since the fluorophore-containing antibody has been applied. We recommend putting a dark cloth, a box, or a tin foil over the Coplin jar, or putting it into a cupboard during incubation.

18. Dilute 1:100 or manufacturer recommended amount as a starting point for optimization for the antibody. Adjustments either up or down of the starting concentration may be necessary to get optimal results.

19. Instead of DAPI, other DNA stains like Sytox Green or Propidium Iodide can be used. A quicker staining can be achieved by using e.g. VectaShield with DAPI included though we get better results with the proposed sequential procedure.

20. Depleting the slide of salts from the PBS buffer might help reduce background fluorescence in DAPI filters.

21. Immunolocalization samples can vary greatly in their outcome. Two samples processed in exactly the same way can result in one that is optimal, having many well-preserved and labeled cells, whereas the second sample looks completely misshapen. Diligently scanning through the bad slide might still be rewarded with a single cell that looks good. Finding cells acceptable for publications may require at least 2 h per slide. As a practical means, we suggest to keep a file record on each type of experiment, noting down the date, any deviation occurring during the preparation, general quality of the sample, and specific outcomes seen and photographed.

22. For more in-depth usage of ImageJ or Fiji, user guides and tutorials are plentiful. We especially recommend one targeted for fluorescence images (*see* ref. 18).

23. Do not hit "apply" after changing Brightness and Contrast because this will permanently change the pixel values as seen when comparing the histograms before and after. Changing pixel values falsifies the information from your image (*see* ref. 18) misrepresenting microscopic observation! Other recommendations to avoid research ethical pitfalls are (1) to always keep the original data with all pertinent metadata (usually recorded by the image acquisition system, if not, diligently note key parameters like filter, magnification, exposure times, gamma adjustment), (2) to work only on copies of the original data, (3) to compare if the edited result still represents something discernible in the original image, and (4) to describe any nonlinear adjustments done, e.g. on the gamma value. Also, never manipulate only specific regions of the image; operations on the whole image are generally fine. One inevitable fail-safe test is to obtain images with the same meaning that stem from independent experiments.

24. For different pixel sizes it is necessary to recalculate scale. If options are not changed, all images taken are usually the same size.

25. Scale bars often cannot be removed once placed. For the purpose of final multipanel figures consider using only one scale bar in the first or last image if all image scales are the same.

26. Do not use the "standard deviation" method since it can render results not matching the original color distribution. We suggest "maximum intensity," but "average intensity," "median," or "sum slices" also work satisfactorily.

27. If adding brightfield, use it as the "gray" option. Another advantage of this image tool is that e.g. different blue tinges from DAPI-stained slides will vanish and all images have the same basic blue.

28. The "Magic Montage" plugin text and installation instructions and a video tutorial can be found at http://imagejdocu.tudor.lu/doku.php?id=video:utilities:creating_montages_with_magic_montage.

# Acknowledgments

We are grateful to Drs. Zac Cande, Wojciech Pawlowski, and Rachel Wang for the maize ASY1 and ZYP1 antisera, and to Chris Lambing for helpful discussion on Fig. 4. This work was supported by the National Science Foundation (IOS: 1025881).

## References

1. Forestan C, Carraro N, Varotto S (2013) Protein immunolocalization in maize tissues. In: De Smet I (ed) Plant organogenesis. Humana Press, New York, pp 207–222

2. de Paula CMP, Techio VH (2014) Immunolocalization of chromosome-associated proteins in plants – principles and applications. Bot Stud 55:1–7

3. Ferdous M, Higgins JD, Osman K et al (2012) Inter-homolog crossing-over and synapsis in Arabidopsis meiosis are dependent on the chromosome axis protein AtASY3. PLoS Genet 8:e1002507

4. Seeliger K, Dukowic-Schulze S, Wurz-Wildersinn R et al (2012) BRCA2 is a mediator of RAD51- and DMC1-facilitated homologous recombination in Arabidopsis thaliana. New Phytol 193:364–375

5. Chelysheva L, Gendrot G, Vezon D et al (2007) Zip4/Spo22 is required for class I CO formation but not for synapsis completion in Arabidopsis thaliana. PLoS Genet 3:e83

6. Higgins JD, Sanchez-Moran E, Armstrong SJ et al (2005) The Arabidopsis synaptonemal complex protein ZYP1 is required for chromosome synapsis and normal fidelity of crossing over. Genes Dev 19:2488–2500

7. Higgins JD, Osman K, Jones GH et al (2014) Factors underlying restricted crossover localization in barley meiosis. Annu Rev Genet 48:29–42

8. Sauer M, Friml J (2010) Immunolocalization of proteins in plants. In: Hennig L, Köhler C (eds) Plant developmental biology. Humana Press, Totowa, pp 253–263

9. Pavlova P, Tessadori F, de Jong HJ et al (2010) Immunocytological analysis of chromatin in isolated nuclei. In: Hennig L, Köhler C (eds) Plant developmental biology. Humana Press, Totowa, pp 413–432

10. Yang X, Yuan L, Makaroff CA (2013) Immunolocalization protocols for visualizing meiotic proteins in Arabidopsis thaliana: method 3. In: Pawlowski WP, Grelon M, Armstrong S (eds) Plant meiosis. Humana Press, New York, pp 109–118

11. Armstrong S, Osman K (2013) Immunolocalization of meiotic proteins in Arabidopsis thaliana: method 2. In: Pawlowski WP, Grelon M, Armstrong S (eds) Plant meiosis. Humana Press, New York, pp 103–107

12. Chelysheva LA, Grandont L, Grelon M (2013) Immunolocalization of meiotic proteins in Brassicaceae: method 1. In: Pawlowski WP, Grelon M, Armstrong S (eds) Plant meiosis. Humana Press, New York, pp 93–101

13. Rossner M, Yamada KM (2004) What's in a picture? The temptation of image manipulation. J Cell Biol 166:11–15

14. Rasband WS (1997) ImageJ. U.S. National Institutes of Health, Bethesda, MD

15. Armstrong SJ, Franklin FCH, Jones GH (2003) A meiotic time-course for Arabidopsis thaliana. Sex Plant Reprod 16:141–149

16. Dukowic-Schulze S, Sundararajan A, Ramaraj T et al (2014) Sequencing-based large-scale genomics approaches with small numbers of isolated maize meiocytes. Plant Genet Genomics 5:57

17. Boateng KA, Yang X, Dong F et al (2008) SWI1 is required for meiotic chromosome remodeling events. Mol Plant 1:620–633

18. Bankhead P (2013) Analyzing fluorescence microscopy images with ImageJ. http://blogs.qub.ac.uk/ccbg/files/2013/06/Analyzing_fluorescence_microscopy_images.pdf

# Chapter 14

# Mapping Recombination Initiation Sites Using Chromatin Immunoprecipitation

## Yan He, Minghui Wang, Qi Sun, and Wojciech P. Pawlowski

## Abstract

Genome-wide maps of recombination sites provide valuable information not only on the recombination pathway itself but also facilitate the understanding of genome dynamics and evolution. Here, we describe a chromatin immunoprecipitation (ChIP) protocol to map the sites of recombination initiation in plants with maize used as an example. ChIP is a method that allows identification of chromosomal sites occupied by specific proteins. Our protocol utilizes RAD51, a protein involved in repair of double-strand breaks (DSBs) that initiate meiotic recombination, to identify DSB formation hotspots. Chromatin is extracted from meiotic flowers, sheared and enriched in fragments bound to RAD51. Genomic location of the protein is then identified by next-generation sequencing. This protocol can also be used in other species of plants, animals, and fungi.

**Key words** Chromosomes, Chromatin, Immunoprecipitation, Antibody, Maize, Recombination, Double-strand breaks (DSBs)

## 1 Introduction

Meiotic recombination is one of the key functions of chromosomes. It is required for faithful segregation of genetic material to the progeny and also generates genetic variation. Recombination is initiated by the formation of double-strand breaks (DSBs) in chromosomal DNA [1, 2]. The breaks are subsequently repaired into either crossovers (COs) or non-crossovers (NCOs), which include gene conversions. DSBs in most species, including plants, vastly outnumber COs [3]. In maize, roughly 500 DSBs are created in each cell during meiosis [4, 5], of which fewer than 20 become CO sites.

Recombination events in most species, including plants, are not uniformly distributed along chromosomes but form distinct hotspots [6–8]. Determining the location of recombination hotspots facilitates the understanding of genome dynamics and evolution as well as the elucidation of factors that cause specific regions of the genome to become recombination hotspots.

Shahryar F. Kianian and Penny M.A. Kianian (eds.), *Plant Cytogenetics: Methods and Protocols*,
Methods in Molecular Biology, vol. 1429, DOI 10.1007/978-1-4939-3622-9_14,
© Springer Science+Business Media New York 2016

COs sites can be inferred by following exchanges of genetic markers (e.g., Single Nucleotide Polymorphisms or SNPs) between parental chromosomes in hybrid progeny [9]. Thus, resolution of CO mapping varies, depending on SNP density, and can be quite poor in regions of limited DNA sequence polymorphism. Sites of meiotic DSBs are most often identified using biochemical approaches [7, 10–12].

One of the most successful methods of identifying DSB hotspot locations genome-wide is finding chromosome sites associated with the RAD51 protein [7, 12] using chromatin immunoprecipitation (ChIP). Following DSB formation, DNA ends flanking DSB sites are resected to form single-stranded DNA ends. These ends become then coated by two recombination proteins, RAD51 and DMC1 [13–16]. RAD51 catalyzes the first step of repair of meiotic DSBs and localizes to DNA segments immediately adjacent to the DSB sites [16]. The protein forms discrete foci on chromosomes during meiotic prophase I (Fig. 1). The number of foci is thought to represent the number of meiotic DSBs.

To conduct ChIP experiments, chromosomal proteins are crosslinked, chromatin is extracted, and enriched in fragments containing the protein of interest using a specific antibody [17, 18]. The antibody-enriched fragments are identified using either whole-genome DNA tiling arrays or next-generation sequencing. The ChIP technique offers several advantages for mapping recombination hotspots. First, all hotspots, not only those that produce COs, can be surveyed. Second, hotspot sites can be determined at a very high resolution of a few hundred base pairs. Third, a very large number of meiocytes can be surveyed in a single experiment. Finally, the mapping resolution is independent of SNP density and hotspots can be mapped in homozygous strains.

In this chapter, we describe a ChIP protocol to map sites of meiotic recombination hotspots in maize using an antibody against RAD51. This protocol is a modification of a previously published general-use ChIP protocol [19] for specific use in mapping

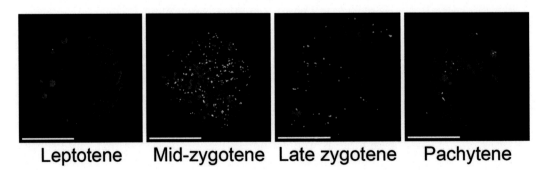

**Leptotene    Mid-zygotene    Late zygotene    Pachytene**

**Fig. 1** The RAD51 protein localize to discrete sites on maize chromosomes during meiotic prophase I. Chromosome sites where RAD51 is located can be determined with high resolution using chromatin immunoprecipitation (ChIP). *Red* = chromatin. *Green* = RAD51. Bar = 10 $\mu$m. Modified from Pawlowski et al. [27]

RAD51-marked DSB sites. The protocol is used by us to map the landscape of DSB hotspots in maize [12]. It can also be used in other species.

## 2    Materials

### 2.1    Reagents

#### 2.1.1    Staging and Collecting Meiotic Flowers

1. Maize plants grown in a controlled environment growth chamber (*see* **Note 1**). We use a 12 h day/12 h night photoperiod, temperature of 31 °C during the day and 22 °C at night, and light intensity of about 600 μmol/m²/s.

2. Acetocarmine stain: 2% acetocarmine powder in 45% acetic acid. Boil the solution for 6–8 h in a flask with boiling stones and an attached reflux column. Then, filter the solution through filter paper when it is still warm. Store stain in a dark bottle at room temperature.

#### 2.1.2    Chromatin Crosslinking

1. Crosslinking buffer: 10 mM Tris–HCl (pH 8.0), 0.4 M sucrose, 10 mM $MgCl_2$, 5 mM β-mercaptoethanol, 1% formaldehyde.

2. 2 M glycine in water.

#### 2.1.3    Chromatin Extraction and Sonication

1. Chromatin extraction buffer A: 10 mM Tris–HCl (pH 8.0), 0.4 M sucrose, 10 mM $MgCl_2$, 1 mM phenylmethylsulfonyl fluoride (PMSF), 5 mM β-mercaptoethanol. Before use, add protease inhibitor (*see* **Note 2**).

2. Chromatin extraction buffer B: 10 mM Tris–HCl (pH 8.0), 0.25 M sucrose, 10 mM $MgCl_2$, 1% Triton X-100, 1 mM phenylmethylsulfonyl fluoride (PMSF), 5 mM β-mercaptoethanol, protease inhibitor (*see* **Note 2**).

3. Chromatin extraction buffer C: 10 mM Tris–HCl (pH 8.0), 1.7 M sucrose, 2 mM $MgCl_2$, 0.15% Triton X-100, 1 mM phenylmethylsulfonyl fluoride (PMSF), 5 mM β-mercaptoethanol, protease inhibitor (*see* **Note 2**).

4. Nuclei lysis buffer: 50 mM Tris–HCl (pH 8.0), 10 mM EDTA, 1% (w/v) SDS, 1 mM phenylmethylsulfonyl fluoride (PMSF), protease inhibitor (*see* **Note 2**).

#### 2.1.4    Chromatin Immunoprecipitation

1. Dynabeads (Invitrogen, Grand Island, NY, USA) (*see* **Note 3**).

2. ChIP dilution buffer: 16.7 mM Tris–HCl (pH 8.0), 1.2 mM EDTA, 167 mM NaCl, 1.1% Triton X-100, 1 mM phenylmethylsulfonyl fluoride (PMSF), protease inhibitor (*see* **Note 2**).

3. Blocking buffer: 16.7 mM Tris–HCl (pH 8.0), 1.2 mM EDTA, 167 mM NaCl, 1.1% Triton X-100, 1 mM phenylmethylsulfonyl fluoride (PMSF), protease inhibitor (*see* **Note 2**).

4. Low salt wash buffer: 20 mM Tris–HCl (pH 8.0), 2 mM EDTA, 150 mM NaCl, 0.1 % SDS, 1 % Triton X-100.

5. High salt wash buffer: 20 mM Tris–HCl (pH 8.0), 2 mM EDTA, 500 mM NaCl, 0.1 % SDS, 1 % Triton X-100.

6. LiCl wash buffer: 10 mM Tris–HCl (pH 8.0), 1 mM EDTA, 250 mM LiCl, 1 % NP-40, 1 % sodium deoxycholate.

7. TE buffer: 10 mM Tris–HCl (pH 8.0), 1 mM EDTA.

8. Elution buffer: 50 mM Tris–HCl (pH 8.0), 10 mM EDTA, 200 mM NaCl, 1 % SDS.

9. 10 mg/mL RNase.

10. 20 mg/mL Proteinase K.

11. PCR purification kit.

12. High-sensitivity DNA quantification kit (*see* **Note 4**).

*2.1.5 ChIP-seq Library Construction and Quality Control*

1. ChIP-seq DNA Sample Prep Kit (Illumina, San Diego, CA, USA) (*see* **Note 5**).

2. PCR purification kit, such as Qiagen's MinElute or QIAquick.

3. Real-Time PCR mix, such as iTaq Universal SYBR Green Supermix (Bio-Rad, Hercules, CA, USA).

4. 10 mg/mL ethidium bromide solution in water.

5. TAE buffer (40 mM Tris, 20 mM acetic acid, 1 mM EDTA, pH 8.0).

6. Gel loading buffer (50 mM Tris pH 8.0, 40 mM EDTA, 40 % (w/v) sucrose) (*see* **Note 6**).

## 2.2 Supplies and Equipment

*2.2.1 Staging and Collecting Meiotic Flowers*

1. Glass scintillation vials or 15 mL plastic tubes to collect flowers for staging.

2. Razor blade.

3. Tweezers with fine tips.

4. Dissecting needle.

5. Rusty nail (*see* **Note 7**).

6. Glass microscope slides and cover slips.

7. Dissecting stereoscope.

8. Bright-field microscope.

*2.2.2 Chromatin Crosslinking*

1. 50 mL conical tubes.

2. Miracloth.

3. Liquid nitrogen.

4. Vacuum desiccator.

| | |
|---|---|
| *2.2.3   Chromatin Extraction and Sonication* | 1. Small ceramic mortar and pestle.<br>2. Liquid nitrogen.<br>3. Miracloth.<br>4. Probe sonicator.<br>5. Microcentrifuge.<br>6. Refrigerated centrifuge.<br>7. Tabletop shaker. |
| *2.2.4   Chromatin Immunoprecipitation* | 1. Magnetic separation stands for bead removal.<br>2. Tube rotator for mixing tube contents. |
| *2.2.5   ChIP-seq Library Construction and Quality Control* | 1. Thermocycler.<br>2. Real-time PCR machine. |

# 3   Methods

*3.1   Staging and Collecting Maize Meiotic Flowers*

1. For DSB mapping, male flowers containing anthers at the zygotene stage of meiotic prophase I should be used (*see* **Note 8**). At this stage, the maize tassel is still inside the stalk. The presence of the tassel can be felt just below the top node of the plant by gently squeezing the leaf whorl. After establishing that the tassel is large enough to be felt, make a small longitudinal incision with a razor blade through the leaves to the tassel, just below the top node.

2. Remove several flowers with needle-nosed forceps. Dissect anthers from the collected flowers on a microscope slide under a stereo dissecting microscope (*see* **Note 9**).

3. Add a drop of acetocarmine solution for staining. Mix anthers with the stain using a dissecting needle or a rusty nail over gentle heat/flame until the color of the stain turns from deep red to purple without boiling the stain solution. Place a cover slip over the anthers and gently press to break the anthers and release meiocytes. Determine the stage of meiosis under a bright-field compound microscope.

4. If the anthers are not yet at the desired meiosis stage, tape over the incision with masking tape and repeat the staging procedure in a day or two.

5. When the tassel is found to contain anthers at the zygotene stage of prophase I, collect the entire plant by cutting it at several nodes below the tassel.

6. Gently remove leaves surrounding the tassel. To prevent the tassel from drying out during dissection, place it on wet paper towels in a tray and put more wet paper towels on top of it.

7. Identify the tassel sections containing anthers at zygotene. Collect individual flowers from at least 20 tassels into 50 mL conical tube (*see* **Note 10**).

**3.2 Chromatin Crosslinking**

1. Add 37 mL of crosslinking buffer (*see* **Note 11**) to the 50 mL conical tube containing the collected flowers. Cap the tube with Miracloth to prevent the tissue from floating on the surface.

2. Vacuum infiltrate the solution for 10 min.

3. Release vacuum slowly and remove Miracloth. Stop the crosslinking reaction by adding 2.5 mL of 2 M glycine. Vacuum infiltrate for 5 min.

4. Decant supernatant and wash the tissue three times with 40 mL of distilled water. After the third wash, dry the tissue between paper towels.

5. Transfer the dry tissue into a new 50 mL conical tube. Snap-freeze in liquid nitrogen and store at −80 °C.

**3.3 Chromatin Extraction and Sonication**

1. Grind the tissue to a fine power with a mortar and pestle in liquid nitrogen.

2. Resuspend the powder in 40 mL of ice-cold chromatin extraction buffer A. Incubate for 20 min at 4 °C with gentle shaking.

3. Filter the solution into a new 50 mL conical tube through two layers of Miracloth placed in a plastic funnel.

4. Centrifuge at $1250 \times g$ for 20 min at 4 °C.

5. Pour out the supernatant and resuspend the pellet in 1 mL of ice-cold extraction buffer B by gently pipetting up and down with a 1000 μL automatic pipette. Transfer the suspension to a 1.5 mL microcentrifuge tube. Incubate on ice for 15 min with occasional agitation.

6. Centrifuge at $20,000 \times g$ in a microcentrifuge for 10 min at 4 °C. Discard the supernatant and resuspend the pellet in 500 μL of ice-cold extraction buffer C by gently pipetting up and down with a 1000 μL automatic pipette (*see* **Note 12**).

7. In a clean 1.5 mL microcentrifuge tube, add 500 μL of extraction buffer C. Layer the resuspended pellet from **step 6** on top of this "cushion."

8. Centrifuge at $20,000 \times g$ in a microcentrifuge for 1 h at 4 °C.

9. Discard the supernatant and resuspend the nuclei pellet in 500 μL of ice-cold nuclei lysis buffer.

10. Sonicate the extracted chromatin on ice using several pulses into fragments of average length of 200–400 bp using

eight sonicator pulses, 5 s each, lasting for 5 s each (*see* **Notes 13** and **14**).

11. Centrifuge the chromatin solution at $20,000 \times g$ in a microcentrifuge for 5 min at 4 °C to pellet tissue debris. Transfer the supernatant containing the chromatin fragments to a new tube.

**3.4    Chromatin Immunoprecipitation**

*3.4.1    Blocking Dynabeads (**See Note 15**)*

1. For each ChIP sample, take 100 μL of Dynabeads slurry into a 1.5 mL microcentrifuge tube.

2. Separate the beads on a magnetic separation stand for 1 min. Without disturbing the beads, pipette out the supernatant.

3. Wash beads twice with 1 mL of ChIP dilution buffer. For each wash, add the buffer and vortex the beads briefly to break coagulates. Then, remove the buffer using the magnet as described in Subheading 3.4.1, **step 2**.

4. Resuspend the beads in 1 mL of blocking buffer. Incubate at 4 °C with gentle shaking for at least 2 h.

5. Wash the beads three times with 1 mL of ChIP dilution buffer as described in Subheading 3.4.1, **step 3**.

6. Add ChIP dilution buffer back to the original bead volume from Subheading 3.4.1, **step 1**.

*3.4.2    Immunoprecipitation and Washes*

1. Take a 10 μL aliquot of the sonicated chromatin sample to use as an input control sample for ChIP product sequencing (*see* **Note 16**).

2. Split the chromatin sample from Subheading 3.3, **step 11** (approx. 450 μL) into three 1.5 mL tubes of equal volume (150 μL in each tube) and dilute the chromatin sample in each tube tenfold by adding 1350 μL of ChIP dilution buffer (*see* **Note 17**).

3. Preclear each chromatin sample by mixing with 40 μL of Dynabeads beads for 3 h with gentle rotation of the tubes on a tube rotator at 4 °C.

4. Separate the beads on the magnetic separation stand.

5. Transfer the supernatant from each tube into a new tube. The first tube will serve as the "no-antibody" control. Add 10 μg of preimmune or normal rabbit IgG to the second tube to use an "IgG control." And 10 μg of your target antibody to the third tube (*see* **Note 18**).

6. Incubate the chromatin samples overnight with rotating at 4 °C.

7. Capture the protein–DNA complexes by adding 40 μL of coated beads and rotating the tubes on a tube rotator for 2.5 h

at 4 °C. Separate beads on the magnetic separation stand and remove the supernatant.

8. Wash the beads five times with 1 mL of each of the following buffers (1) low salt wash buffer, (2) high salt wash buffer, (3) LiCl wash buffer, and twice in TE buffer. To conduct the washes, rotate the tubes for 5 min at 4 °C and remove the buffer as described in Subheading 3.4.1, **step 2**. After the final wash, make sure to remove all TE.

9. Add 200 μL of freshly prepared elution buffer. Resuspend beads by vortexing and incubate at room temperature for 30 min with occasional agitation.

10. Centrifuge at $2000 \times g$ for 1 min and collect the supernatant into a new tube.

11. Conduct a second immunoprecipitation (*see* **Note 19**) by repeating **steps 5–8** of Subheading 3.4.2.

12. Add 200 μL of freshly prepared elution buffer. Resuspend beads by vortexing and incubate at 65 °C for 30 min with occasional agitation.

13. Centrifuge at $2000 \times g$ for 1 min and collect the supernatant into a new tube.

*3.4.3 Decrosslinking*

1. Add 4 μL of 10 mg/mL RNase to the supernatants from Subheading 3.4.2, **step 13**, and incubate at 37 °C for 1.5 h.

2. Add 4 μL of 20 mg/mL Proteinase K and incubate at 45 °C for 2 h.

3. Reverse crosslink at 65 °C for 8 h or overnight.

*3.4.4 DNA Recovery*

1. Purify DNA from each sample from Subheading 3.4.3, **step 3** using a PCR purification kit, eluting in 30 μL of $H_2O$.

2. Measure the DNA concentration with a fluorometer following manufacturer's instructions.

**3.5 ChIP-seq Library Construction and Sequencing**

1. Follow the current Illumina protocol to construct a sequencing library using the ChIP-seq DNA Sample Prep Kit (Illumina, San Diego, CA, USA) with the exception that DNA size selection should be done after the PCR step [7].

2. Perform a quality control experiment to validate the sequencing library. To do this, design a pair of PCR primers for a genome region known to be a recombination hotspot and a pair of primers for a random region, such as the *Ubiquitin* locus. Conduct three independent quantitative PCR reactions using the iTaq Universal SYBR Green Supermix (Bio-Rad, Hercules, CA, USA) following manufacturer's instructions. Use an average of the three experiments to calculate hotspots enrichment. First, normalize the ChIP data using enrichment

of a known hotspot region in the ChIP sample to that in the input sample. Then, normalize to the *Ubiquitin* gene region using the following equation: $2^{[Ct(Hotspot\ region\_ChIP)-Ct(Hotspot\ region\_Input)]}/2^{[Ct(Ubiquitin\ region\_ChIP)-Ct(Ubiquitin\ region\_Input)]}$.

**3.6 Computational Analyses**

*3.6.1 Processing and Mapping Illumina Reads to the Genome Scaffold*

1. Perform base calling and read quality control using the standard Illumina protocol.

2. Align reads that passed quality control to the reference genome sequence using the Burrows-Wheeler Aligner (BWA) [20].

3. Trim reads progressively at the 3′ termini 1 bp at a time until they can be mapped to the genome scaffold with no more than two mismatches. Only reads that are longer than 40 bp after trimming should be aligned.

4. To identify recombination hotspots, conduct a peak detection analysis using MACS [21]. We use the following parameters for RAD51 ChIP peak detection: bandwidth = 800 bp, shift size = 400 bp, MACS mode off, and *q*-value cutoff = 0.01.

5. Use several control datasets to identify regions of RAD51 ChIP enrichment, such as: (1) input chromatin (from Subheading 3.4.2, **step 1**), (2) ChIP conducted using preimmune or normal rabbit IgG on meiotic chromatin (from Subheading 3.4.2, **step 5**), and (3) ChIP conducted with the anti-RAD51 antibody using chromatin extracted from young seedlings.

6. The Integrative Genomics Viewer (IGV) [22] can be used to visualize high-resolution DSB hotspot maps.

# 4    Notes

1. Plants for DSB mapping experiments should be grown in controlled environment growth chambers as temperature is known to affect both the number and distribution of recombination events (*see* ref. 23, 24).

2. We use one tablet of cOmplete Protease Inhibitor (Roche Applied Science, Indianapolis, IN, USA) per 50 mL of buffer. However, similar products from other manufacturers can be used instead.

3. Similar products from other manufacturers can be used instead but should be tested in pilot experiments. Use magnetic beads coupled to Protein A or Protein G, depending on the animal species in which the antibody of choice was produced.

4. We use Quant-IT dsDNA HS Assay Kit (Invitrogen, Grand Island, NY, USA). However, similar products from other manufacturers can be used instead.

5. Equivalent kits from other manufacturers, such as NEB (Ipswich, MA, USA) or homemade kits can also be used.

6. Any homemade or commercially available gel loading buffer can be used.

7. Iron oxide that leaches from the rusty nail enhances the staining reaction. Too little iron oxide will result in weak staining.

8. DSBs in plants are generated very early in meiosis, most likely before the onset of leptotene (*see* ref. 25, 26). However, RAD51 foci are present on chromosomes from late leptotene to mid-pachytene but exhibit their peak at mid-zygotene (*see* ref. 4, 27).

9. If flower samples need to be transported, they can be collected into glass scintillation vials or tubes containing Farmer's fixative (three volumes of 100 % ethanol, one volume of glacial acetic acid).

10. Using anthers instead of whole flowers as input material might help reduce experimental background but anthers are much more time consuming to collect than whole flowers.

11. The length of crosslinking and the formaldehyde concentration need to be optimized for each tissue type. Insufficient crosslinking may result in decreased binding of the ChIP antibody while excessive crosslinking may lead to nonspecific binding.

12. Avoid introducing air bubbles or forming froth on the surface as this may lead to protein degradation in subsequent steps.

13. To produce desired fragment sizes, sonication conditions need to be optimized for every sonicator type and tissue type. Over-sonication will lead to DNA degradation, whereas insufficient sonication will lead to nonspecific antibody binding and decrease ChIP yield.

14. Heat generated during sonication may cause protein degradation. To avoid it, keep samples on ice during the entire procedure and allow at least 30 s between each sonicator pulse to let the samples cool down.

15. The bead blocking step can be carried out before starting the ChIP experiment. After blocking, beads can be stored at 4 °C. Blocking the beads decreases nonspecific binding of the antibody. We strongly recommend including this step, even though it is not always suggested in published ChIP protocols. Do not use DNA as a blocking reagent if the ChIP DNA product will be analyzed by sequencing. Otherwise, most of the sequence reads will represent carrier DNA.

16. The sonicated chromatin sample needs to be decrosslinked before DNA extraction. To do this, add 140 μL of TE buffer, 5 μL of 5 M NaCl, and 10 μL of 10 % SDS to a 10 μL aliquot

of sonicated chromatin. Reverse crosslink overnight at 65 °C. Purify DNA using a PCR purification kit. To check sonication efficiency, electrophorese an aliquot of the extracted DNA in 2 % agarose gel.

17. From this step on, use low-retention microcentrifuge tubes.

18. Using antibodies with high specificity and affinity to native proteins is very important. Performance of an antibody in immunolocalization and/or western blot experiments may not be a predictor of its suitability for ChIP. ChIP experiments, with appropriate negative controls, should be carried out to determine antibody's performance. The amount of antibody that should be used in a ChIP experiment depends on the affinity between the antibody and the antigen, which varies from one antibody to another, and should be optimized for each antibody.

19. The second round of immunoprecipitation is used to increase specificity. It should be performed in the same manner as the first immunoprecipitation, except the final elution step, which is carried out at 65 °C.

## Acknowledgment

Research to develop this protocol was supported by a grant from National Science Foundation (IOS-1025881) to WPP.

## References

1. Keeney S, Giroux CN, Kleckner N (1997) Meiosis-specific DNA double-strand breaks are catalyzed by Spo11, a member of a widely conserved protein family. Cell 88:375–384

2. Grelon M, Vezon D, Gendrot G, Pelletier G (2001) AtSPO11-1 is necessary for efficient meiotic recombination in plants. EMBO J 20:589–600

3. Mezard C, Vignard J, Drouaud J, Mercier R (2007) The road to crossovers: plants have their say. Trends Genet 23:91–99

4. Franklin AE, McElver J, Sunjevaric I, Rothstein R, Bowen B, Cande WZ (1999) Three-dimensional microscopy of the Rad51 recombination protein during meiotic prophase. Plant Cell 11:809–824

5. Pawlowski WP, Golubovskaya IN, Cande WZ (2003) Altered nuclear distribution of recombination protein RAD51 in maize mutants suggests the involvement of RAD51 in meiotic homology recognition. Plant Cell 15:1807–1816

6. Mezard C (2006) Meiotic recombination hotspots in plants. Biochem Soc Trans 34:531–534

7. Smagulova F, Gregoretti IV, Brick K, Khil P, Camerini-Otero RD, Petukhova GV (2011) Genome-wide analysis reveals novel molecular features of mouse recombination hotspots. Nature 472:375–378

8. Drouaud J, Camilleri C, Bourguignon PY, Canaguier A, Berard A, Vezon D et al (2006) Variation in crossing-over rates across chromosome 4 of *Arabidopsis thaliana* reveals the presence of meiotic recombination "hot spots". Genome Res 16:106–114

9. Li X, Li L, Yan J (2015) Dissecting meiotic recombination based on tetrad analysis by single-microspore sequencing in maize. Nat Commun 6:6648

10. Pratto F, Brick K, Khil P, Smagulova F, Petukhova GV, Camerini-Otero RD (2014) DNA recombination. Recombination initiation maps of individual human genomes. Science 346:1256442

11. Prieler S, Penkner A, Borde V, Klein F (2005) The control of Spo11's interaction with meiotic recombination hotspots. Genes Dev 19:255–269

12. He Y, Wang M, Dukowic-Schulze S, Bradbury P, Eichten S, Sidhu GK et al (2016) Genomic features shaping the landscape of meiotic double strand break hotspots in maize. Submitted

13. Bishop DK, Park D, Xu L, Kleckner N (1992) DMC1: A meiosis-specific yeast homolog of *Escherichia coli* recA required for recombination, synaptonemal complex formation, and cell cycle progression. Cell 69:439–456

14. Masson J-Y, West SC (2001) The Rad51 and Dmc1 recombinases: a non-identical twin relationship. Trends Biochem Sci 26:131–136

15. Shibata T, Nishinaka T, Mikawa T, Aihara H, Kurumizaka H, Yokoyama S et al (2001) Homologous genetic recombination as an intrinsic dynamic property of a DNA structure induced by RecA/Rad51-family proteins: a possible advantage of DNA over RNA as genomic material. Proc Natl Acad Sci U S A 98:8425–8432

16. Kurzbauer MT, Uanschou C, Chen D, Schlogelhofer P (2012) The recombinases DMC1 and RAD51 are functionally and spatially separated during meiosis in Arabidopsis. Plant Cell 24:2058–2070

17. O'Neill LP, Turner BM (1996) Immunoprecipitation of chromatin. Methods Enzymol 274:189–197

18. Collas P (2010) The current state of chromatin immunoprecipitation. Mol Biotechnol 45:87–100

19. He Y, Sidhu GK, Pawlowski WP (2013) Chromatin immunoprecipitation for studying chromosomal localization of meiotic proteins in maize. Methods Mol Biol 990:191–201

20. Li H, Durbin R (2009) Fast and accurate short read alignment with Burrows-Wheeler transform. Bioinformatics 25:1754–1760

21. Zhang Y, Liu T, Meyer CA, Eeckhoute J, Johnson DS, Bernstein BE et al (2008) Model-based analysis of ChIP-Seq (MACS). Genome Biol 9:R137

22. Thorvaldsdottir H, Robinson JT, Mesirov JP (2013) Integrative Genomics Viewer (IGV): high-performance genomics data visualization and exploration. Brief Bioinform 14:178–192

23. Francis KE, Lam SY, Harrison BD, Bey AL, Berchowitz LE, Copenhaver GP (2007) Pollen tetrad-based visual assay for meiotic recombination in Arabidopsis. Proc Natl Acad Sci U S A 104:3913–3918

24. Higgins JD, Perry RM, Barakate A, Ramsay L, Waugh R, Halpin C et al (2012) Spatiotemporal asymmetry of the meiotic program underlies the predominantly distal distribution of meiotic crossovers in barley. Plant Cell 24:4096–4109

25. Sanchez-Moran E, Santos JL, Jones GH, Franklin FC (2007) ASY1 mediates AtDMC1-dependent interhomolog recombination during meiosis in Arabidopsis. Genes Dev 21:2220–2233

26. Pawlowski WP, Golubovskaya IN, Timofejeva L, Meeley RB, Sheridan WF, Cande WZ (2004) Coordination of meiotic recombination, pairing, and synapsis by PHS1. Science 303:89–92

27. Pawlowski WP, Golubovskaya IN, Cande WZ (2003) Altered nuclear distribution of recombination protein RAD51 in maize mutants suggests involvement of RAD51 in the meiotic homology recognition. Plant Cell 8:1807–1816

# Chapter 15

## Chromatin Immunoprecipitation to Study The Plant Epigenome

### Zidian Xie and Gernot Presting

### Abstract

Chromatin immunoprecipitation (ChIP) has been widely used for studying in vivo protein–DNA interactions for decades. ChIP is a powerful tool that is adaptable for studying epigenetic modifications at certain genomic loci or the genomic level. Given its utility in studying the epigenome and the many technical challenges, we present a detailed in-lab ChIP protocol primarily used for studying histone modifications in plants, but can be easily adapted for other chromatin targets in other species.

**Key words** Chromatin immunoprecipitation, Epigenome, Epigenetic modifications, Histone, Cross-linking, Antibody

## 1 Introduction

Many aspects of plant development, including gametogenesis, seed development, and flowering time, are directly or indirectly regulated by epigenetic marks [1]. As sessile organisms, plants developed effective strategies to rapidly respond to environmental changes, mainly through epigenetic modifications. Epigenetic modifications, including DNA methylation and histone modification, regulate gene expression by altering chromatin structure. In addition, at key locations such as active genes and centromeres, canonical histones are replaced by histone variants such as H2A.Z and CENH3, which influence local chromatin structure and gene activity [2–4]. Given the significance of epigenetic modifications in plants, research focus has shifted to studying genome-wide epigenetic modifications—the epigenome [5, 6]. To study the epigenome, and in particular histone modifications, the most powerful approach is chromatin immunoprecipitation coupled with tiling microarray (ChIP-chip) or deep sequencing (ChIP sequencing), which reveals epigenetic modifications at the genomic level.

Shahryar F. Kianian and Penny M.A. Kianian (eds.), *Plant Cytogenetics: Methods and Protocols*,
Methods in Molecular Biology, vol. 1429, DOI 10.1007/978-1-4939-3622-9_15,
© Springer Science+Business Media New York 2016

With the increasing interest in the epigenome, a number of chromatin immunoprecipitation (ChIP) techniques have been developed and optimized for studying genome-wide histone modifications [7–12]. In general, the ChIP technique is considered technically challenging. The two most critical steps that determine the success of ChIP are chromatin isolation and the antibody selection. While the chromatin is relatively stable, the interaction in the chromatin complex is compromised if harsh conditions (for example, SDS buffer and over-digestion of chromatin) are used for chromatin isolation. To prevent disassembly of the chromatin complex, cross-linking with a reagent such as formaldehyde is necessary [13]. Performing uniform cross-linking is essential to the ChIP technique as the efficiency of cross-linking varies significantly for different plant material. In addition, how much cross-linking reagent should be used and how long the cross-linking is allowed to proceed are other issues for consideration and optimization. Another important step is the antibody selection, which might be the most critical factor to ensure successful ChIP. In general, polyclonal antibodies have relatively high binding affinity but low specificity, while monoclonal antibodies have relatively high specificity but low affinity. The quality of antibodies used in ChIP is crucial. Here, we present a detailed ChIP protocol developed from previously reported ChIP protocols [11, 12]. This modified protocol includes extensive discussion on some critical steps so that it can be easily followed and adapted to individual needs.

## 2    Materials

1. Immunoprecipitation magnetic beads (such as Protein A-Dynabeads® from Invitrogen).

2. Micrococcal Nuclease, MNase dissolved in 50 % filter-sterilized glycerol with a final concentration of 15 U/μl, and kept at −20 °C.

3. 1.5 ml graduated, low-retention microcentrifuge tubes.

4. Miracloth.

5. Glycogen solution at 20 mg/ml of molecular biology grade glycogen free of DNAses and RNases.

6. RNase A at 10 mg/ml.

7. Proteinase K at 10 mg/ml.

8. 100 mM PMSF solution: 174.2 mg Phenylmethylsulfonyl fluoride dissolved in 10 ml 100 % ethanol and stored in −20 °C (protect from light).

9. **Cross-linking buffer**: 0.4 M Sucrose, 10 mM Tris–HCl, pH 8.0, 1 mM EDTA, 1% Formaldehyde, 0.1 mM Phenylmethylsulfonyl fluoride (PMSF).

10. **M1 buffer**: 11.9 % Hexylene glycol, 10 mM $KPO_4$, pH 7.0, 100 mM NaCl, 5 mM beta-mercaptoethanol, 0.1 mM PMSF (freshly added), 1× Plant protease inhibitor cocktail (freshly added).

11. **M2 buffer**: 8.85 % Hexylene glycol, 10 mM $KPO_4$, pH 7.0, 10 mM $MgCl_2$, 0.5 % Triton X-100, 5 mM beta-mercaptoethanol, 100 mM NaCl.

12. **Incubation buffer**: 20 mM Tris–HCl, pH 6.8, 50 mM NaCl, 5 mM EDTA, 0.1 mM PMSF (freshly added), 1× Plant protease inhibitor cocktail (freshly added).

13. **MNB buffer**: 50 mM Tris–HCl, pH 8.0, 2.5 mM $CaCl_2$, 4 mM $MgCl_2$, 0.3 M Sucrose.

14. **PK digestion buffer (10×)**: 50 mM EDTA, 100 mM Tris–HCl, pH 7.4, 0.25 % SDS.

15. **Low salt washing buffer**: 150 mM NaCl, 0.1 % SDS, 1 % Triton X-100, 2 mM EDTA, 20 mM Tris–HCl, pH 8.0, 0.1 mM PMSF (freshly added).

16. **High salt washing buffer**: 500 mM NaCl, 0.1 % SDS, 1 % Triton X-100, 2 mM EDTA, 20 mM Tris–HCl, pH 8.0, 0.1 mM PMSF (freshly added).

17. **LiCl washing buffer**: 0.25 M LiCl, 1 % NP-40, 1 % Sodium deoxycholate, 1 mM EDTA, 10 mM Tris–HCl, pH 8.0, 0.1 mM PMSF (freshly added).

18. **TE buffer**: 10 mM Tris–HCl, pH 8.0, 1 mM EDTA, 0.1 mM PMSF (freshly added).

19. **Elution buffer**: 1 % SDS, 0.1 M $NaHCO_3$.

# 3    Methods

### 3.1    Chromatin Preparation

1. Grind 2 g of plant material of interest (for example, maize young ears or leaves, soybean seeds) in liquid nitrogen using mortar and pestle (*see* **Notes 1** and **2**).

2. Transfer the powder into 50 ml tubes, add 30 ml of freshly made cross-linking buffer containing 1 % Formaldehyde and vortex briefly.

3. Incubate on ice for 20 min, mix by gently shaking every 5 min (*see* **Notes 3** and **4**).

4. Add glycine to a final concentration of 0.1 M to stop cross-linking, mix by gentle shaking. Incubate on ice for another 5 min.

5. Filter the cross-linked chromatin through two layers of Miracloth to remove plant debris (*see* **Note 5**).

6. Spin down at $1000 \times g$ at 4 °C for 10 min, after transferring the filtered chromatin solution into a number of 1.5 ml low-retention microcentrifuge tubes (*see* **Note 6**).

7. After carefully removing the supernatant, resuspend the pellet (containing crude chromatin) by pipetting up and down gently in 1 ml of ice-cold M1 buffer and spin at $1000 \times g$ at 4 °C for 10 min.

8. Repeat **step 7**.

9. Combine all tubes of a single sample into one tube, and wash one more time with 1 ml of ice-cold M2 buffer.

10. Add 400 μl MNB buffer and the desired amount of MNase and incubate at 37 °C for the desired time (*see* **Notes 7** and **8**).

11. Stop MNase digestion by adding 1/10 volume of 0.5 M EDTA and keep on ice.

12. Spin down at $16,000 \times g$ in a microcentrifuge at 4 °C for 10 min.

13. Save the supernatant containing the chromatin, resuspend the pellet with 400 μl of incubation buffer, and incubate on ice for 60 min to extract more chromatin from the pellet.

14. Spin down at $16,000 \times g$ at 4 °C for 10 min, and combine two supernatant fractions, which contain the digested chromatin.

**3.2 Isolation of Specific Chromatin–DNA Complex**

1. Add 40 μl of prewashed immunoprecipitation magnetic beads in incubation buffer and 2 μl of IgG antibodies to the chromatin solution to preclear the chromatin, and rotate/mix at 4 °C for at least 2 h (*see* **Notes 9** and **10**).

2. Use a magnetic stand to remove the beads, save 1/10 of the supernatant as an input control fraction for later steps (keep at −20 °C).

3. Divide the rest of the supernatant into several low-retention tubes (*see* **Note 11**), and add a certain amount of antibodies into different tubes (*see* **Notes 12** and **13**), as well as 500 μl of incubation buffer (*see* **Note 14**). Rotate at 4 °C overnight.

4. Add 20 μl of prewashed immunoprecipitation magnetic beads in incubation buffer into each tube and continue to rotate at 4 °C for another 3–5 h.

5. Prepare all the washing buffers (keep at 4 °C) and make FRESH elution buffer.

6. Do washes in the following order: Low salt washing buffer first, followed by high salt washing buffer and finally the LiCl washing buffer. Wash twice with each buffer using 1 ml of buffer each time. For each wash do the following: briefly spin to bring the buffer down to the bottom of tubes ($1000 \times g$ in microcentrifuge at 4 °C for 10 s), place the tubes into the magnetic stand for 1 min, carefully remove the supernatant without touching the beads, put new buffer in, rotate for 5 min or longer at 4 °C.

7. Following the final LiCl wash, wash the beads with 1 ml of ice-cold TE buffer. Remove TE buffer after 1 min in the magnetic stand and add 200 μl of elution buffer. Invert to mix and incubate at 65 °C for 15 min or longer, shaking from time to time.

8. Save the supernatant to a new 1.5 ml tube after the beads attach to the magnetic stand, add another 200 μl of elution buffer to the beads and incubate again at 65 °C for 15 min (*see* **Note 15**). Combine the two supernatants, which should contain the eluted chromatin target complexes.

9. Add 16 μl of 5 M NaCl to each tube and incubate at 65 °C overnight. In addition, take out 10 % of the chromatin used for each ChIP reaction from the previously saved input fraction (**step 2** of Subheading 3.2), bring the volume up to 400 μl with elution buffer, add 16 μl of 5 M NaCl, and incubate at 65 °C overnight together with the rest of samples.

**3.3 DNA Isolation**

1. Add 1 μl of RNase A (10 mg/ml) and incubate at 37 °C for 30 min.

2. Add 1/10 volume of 10× PK digestion buffer and 1.5 μl of 10 mg/ml Proteinase K to each tube, and incubate at 45 °C for at least 1 h.

3. Add equal volume of phenol:chloroform:ispropanol (25:24:1), vortex for 1 min and spin at $16,000 \times g$ at room temperature for 10 min.

4. Carefully save the supernatant into a new tube and add an equal volume of chloroform, vortex for 1 min and spin at 16,000 x g in microcentrifuge at room temperature for 10 min.

5. Save the supernatant into new tubes, and add 1/10 volume of 3 M sodium acetate (pH 5.2), two volumes of 100 % ethanol and 0.2 μl of glycogen. Leave at −20 °C for at least 2 h (preferably overnight).

6. Spin at $16,000 \times g$ at 4 °C for 30 min, wash the pellet with 1 ml of ice-cold 80 % ethanol, and spin at $16,000 \times g$ at 4 °C for 10 min. Carefully remove as much residual ethanol as possible without disturbing the pellet.

7. Air-dry the pellet in the sterile hood (leave the tubes open) for 10–15 min, the pellet should have a white color but can sometimes be invisible. After drying, add 20 μl of TE buffer to dissolve DNA (*see* **Note 16**).

8. Measure DNA concentration (*see* **Note 17**).

9. ChIPed DNA can be used for real-time PCR to verify enrichment for specific genomic DNA regions (*see* **Note 18**), and then used for deep sequencing, such as Illumina sequencing (*see* **Note 19**).

## 4   Notes

1. The amount of plant material varies from 500 mg to 2 g, or can be scaled up, depending on how many ChIP reactions you want to perform. In general, 250 mg plant tissue should be suitable for one ChIP reaction.

2. Due to different properties of plant materials, it is very hard to do uniform cross-linking with intact plant tissues or organs, such as roots and seeds. Therefore, it becomes much easier to do cross-linking with fine plant tissue powder, which can lead to uniform cross-linking. More importantly, the optimized conditions for cross-linking with fine tissue powder can be applied to almost any plant material. Thus, it is critical to grind plant material to a very fine powder. The better samples are ground, the more uniform the cross-linking will be and the more chromatin one will get.

3. It is a good idea to do cross-linking for different plant samples at the same time to reduce the ChIP variation between replicates due to the cross-linking difference.

4. The cross-linking conditions (such as concentration of formaldehyde and cross-linking time) must be optimized. Less cross-linking usually results in instability of chromatin during ChIP procedures and leads to low DNA yield. Without reverse cross-linking, the chromatin with sufficient cross-linking will yield very little or no DNA, which can be used as a criterion to determine whether the cross-linking is sufficient. On the other hand, excessive cross-linking will result in low DNA yield even after reverse cross-linking. These criteria can be used to optimize cross-linking conditions. In our hands, with 1 % formaldehyde 20 min on ice is the best condition for cross-linking.

5. Sometimes you need to squeeze the Miracloth gently in order to get most of the solution out. However, do not let the debris get into the flow-through.

6. Low-retention tubes are preferred for all steps in the ChIP procedure, since any chromatin and DNA attached to the tubes will reduce ChIP yield. In addition, filtered tips are recommended to reduce cross-contamination between samples and ChIP reactions.

7. The common approaches to shear the chromatin include physical sonication using a bioruptor (Diagenode) or sonicator, and enzymatic digestion using MNase. For the purpose of epigenomic studies, MNase digestion is preferred since it provides information about nucleosome positioning.

8. The amount of MNase depends on the source of MNase and the chromatin amount. The condition for MNase digestion

can be optimized by running DNA from digested chromatin on a 1% agarose gel with EtBr staining. For epigenomic studies, the length for the majority of isolated chromatin should be one to a few nucleosomes. In our hands, for the chromatin from 0.5 g of plant powder, 4.5 units of MNase (USB) at 37 °C for 20 min will digest most of chromatin into mononucleosomes.

9. To achieve better ChIP performance, this preclearing step is important, since it will remove most of the chromatin that are nonspecifically attached to the immunoprecipitation magnetic beads (e.g., Dynabeads) or the antibody. If the corresponding preimmune serum for the antibody is available, it should be used. Otherwise, for commercial antibodies, IgG can be used instead.

10. Although salmon sperm/protein A-agarose is another choice, protein A-Dynabead is preferred since the residual salmon sperm DNA after ChIP can be detected by deep sequencing, which will introduce undesired sequencing noise.

11. The number of tubes depends on the number of ChIP reactions and replicates that will be performed.

12. Many commercial antibodies recognizing histones and histone modifications designed for ChIP are available, such as antibodies from Abcam and Millipore, and most of them can be used for plants. We have successfully used antibodies against unmodified histone H3 (Abcam, ab1791), as well as histone H3 modified with trimethyl K4 (Abcam, ab8580), dimethyl K9 (Millipore, 07-212), and trimethyl K27 (Millipore, 07-449) for soybean ChIP. As a negative control, IgG antibody should be included.

13. The amount of each antibody used for ChIP should be optimized based on ChIP enrichment, which might be different for different antibodies. In our hands, the amount of each histone antibody used is 1~4 μg/ChIP.

14. To allow effective mixing of the solution in the tubes, another 500 μl of incubation buffer should be added into each tube.

15. The incubation time can be longer, such as 1 h.

16. Incubation at 37 °C for 20 min should help remove excess solution.

17. Due to extreme low yield of ChIP procedure, DNA quantification using standard Nanodrop is not accurate. Instead, bioanalyzer or Nanodrop coupled with picogreen staining should be used to check DNA concentration.

18. Real-time PCR should follow standard real-time PCR protocol with some minor modification depending on primers. To compute ChIP fold enrichment, input DNA should be

included. A DNA region that is well known to be a target of the desired epigenetic modifications (targeted DNA) or not to be a target for the same epigenetic modifications (reference DNA) should be used to calculate fold enrichment. As another negative control, the IgG control sample should not give any ChIP enrichment. The formula to calculate ChIP fold enrichment is $(\text{ChIPedDNA}_{\text{target}}/\text{InputDNA}_{\text{target}})/(\text{ChIPedDNA}_{\text{reference}}/\text{InputDNA}_{\text{reference}})$.

19. Given the low DNA yield from ChIP procedure, DNA needs to be amplified prior to deep sequencing. One can use the GenomePlex Whole Genome Amplification (WGA) kit (Sigma) to perform DNA amplification.

## Acknowledgment

The University of Hawaii funded this project.

## References

1. Feng S, Jacobsen SE (2011) Epigenetic modifications in plants: an evolutionary perspective. Curr Opin Plant Biol 14:179–186

2. Brickner DG et al (2007) H2A.Z-mediated localization of genes at the nuclear periphery confers epigenetic memory of previous transcriptional state. PLoS Biol 5:e81

3. Kumar SV, Wigge PA (2010) H2A.Z-containing nucleosomes mediate the thermosensory response in Arabidopsis. Cell 140:136–147

4. Zhang W, Lee HR, Koo DH, Jiang J (2008) Epigenetic modification of centromeric chromatin: hypomethylation of DNA sequences in the CENH3-associated chromatin in Arabidopsis thaliana and maize. Plant Cell 20:25–34

5. Zhang X et al (2006) Genome-wide high-resolution mapping and functional analysis of DNA methylation in arabidopsis. Cell 126:1189–1201

6. Zilberman D, Gehring M, Tran RK, Ballinger T, Henikoff S (2007) Genome-wide analysis of Arabidopsis thaliana DNA methylation uncovers an interdependence between methylation and transcription. Nat Genet 39:61–69

7. Saleh A, Alvarez-Venegas R, Avramova Z (2008) An efficient chromatin immunoprecipitation (ChIP) protocol for studying histone modifications in Arabidopsis plants. Nat Protoc 3:1018–1025

8. Lin X, Tirichine L, Bowler C (2012) Protocol: chromatin immunoprecipitation (ChIP) methodology to investigate histone modifications in two model diatom species. Plant Methods 8:48

9. Ricardi MM, Gonzalez RM, Iusem ND (2010) Protocol: fine-tuning of a chromatin immunoprecipitation (ChIP) protocol in tomato. Plant Methods 6:11

10. Wagschal A, Delaval K, Pannetier M, Arnaud P, Feil R (2007) Chromatin immunoprecipitation (ChIP) on unfixed chromatin from cells and tissues to analyze histone modifications. CSH Protoc 3, doi: 10.1101/pdb.prot4767

11. Wang H, Tang W, Zhu C, Perry SE (2002) A chromatin immunoprecipitation (ChIP) approach to isolate genes regulated by AGL15, a MADS domain protein that preferentially accumulates in embryos. Plant J 32:831–843

12. Johnson L, Cao X, Jacobsen S (2002) Interplay between two epigenetic marks. DNA methylation and histone H3 lysine 9 methylation. Curr Biol 12:1360–1367

13. Morohashi K, Xie Z, Grotewold E (2009) Gene-specific and genome-wide ChIP approaches to study plant transcriptional networks. Methods Mol Biol 553:3–12

# INDEX

Shahryar F. Kianian and Penny M.A. Kianian (eds.), *Plant Cytogenetics: Methods and Protocols*,
Methods in Molecular Biology, vol. 1429, DOI 10.1007/978-1-4939-3622-9,
© Springer Science+Business Media New York 2016

Printed in the United States
By Bookmasters